Cartographic Design
Using ArcView® GIS

Ed Madej

OnWord Press
Thomson Learning™

Africa • Australia • Canada • Denmark • Japan • Mexico • New Zealand
Philippines • Puerto Rico • Singapore • United Kingdom • United States

Trademarks

ArcView and ArcInfo are registered trademarks of the Environmental Systems Research Institute (ESRI), Inc. Avenue is an ESRI trademark.

OnWord Press Staff

Publisher: Alar Elken

Executive Editor: Sandy Clark

Acquisitions Editor: James Gish

Managing Editor: Carol Leyba

Development Editor: Daril Bentley

Editorial Assistant: Fionnuala McAvey

Executive Marketing Manager: Maura Theriault

Executive Production Manager: Mary Ellen Black

Production and Art & Design Coordinator: Cynthia Welch and Leyba Associates

Manufacturing Director: Andrew Crouth

Technology Project Manager: Tom Smith

Cover Design by Cammi Noah

Copyright © 2000 by OnWord Press
SAN 694-0269
First edition 2000
10 9 8 7 6 5 4 3
Printed in Canada

Library of Congress Cataloging-in-Publication Data
Madej, Ed.
    Cartographic design using ArcView GIS / Ed Madej.
        p. cm.
    Includes index.
    ISBN 1-56690-187-1
    1. Cartography—Data processing. 2. ArcView. I. Title.

GA102.4.E4 M32 1999
526'0285—dc21   99-058421
                                                       CIP

For more information, contact
OnWord Press   An imprint of Thomson Learning
Box 15-015   Albany, New York USA 12212-15015

About the Author

Ed Madej is a GIS programmer, ESRI authorized ArcView instructor, and cartographer in private practice in Helena, Montana. He has been producing maps as a cartographer since 1980 and has been a GIS instructor since 1994. His maps have been published in two dozen books, plus many newspapers and magazines, ranging from *The New York Times* to *Sierra* magazine. He has just finished work on the first digitally produced highway travel map for the state of Montana, to be released in November 2000.

Acknowledgments

Most books are group efforts and *Cartographic Design Using ArcView GIS* is no exception. A hearty thanks to my editors at OnWord Press Daril Bentley and Carol Leyba for their invaluable advice and endless patience (OK, almost endless). Graphic designer Cammi Noah helped turn the tortuous color insert into reality, and production editor Cynthia Welch processed many of the illustrations.

My reviewers, fellow ArcView zen masters Lydia Bailey in Kalispell and Dave Highness in Helena, provided more than just technical advice; they provided much-needed encouragement throughout the long writing process. Many of my computer colleagues were good sounding boards for portions of the manuscript: Fred Gifford, Allan Cox, and Kris Larson for many of the GIS tips; Velda Welch for help with Internet maps; and James Conner for printing and color reproduction help. Yes, Jim, Macs still do better color and font management than Wintel boxes!

Professor Paul Wilson at the University of Montana gave great advice for cartographic reference material. I would be remiss in not thanking many of my hundreds of ArcView students, whose questions provided the idea of writing a book in the first place.

It is a poor carpenter (or cartographer) that curses his tools, and any criticisms of ESRI's ArcView are meant in a constructive manner. It is still the best GIS program around and I am sure it will get even better with version 4.0. I would like to thank the many members of the ArcView software team whose brains I have picked at the annual ESRI user's convention. Thanks to ArcView master Michael Blongiewicz at ESRI, who got me started with much of my GIS education. Catherine McCoy at ESRI Olympia generously provided trial copies of ArcPress for ArcView, and several other ArcView extensions that made writing several of the chapters possible.

I have tried to acknowledge in the text the dozens of private ArcView developers who produce incredible extensions, often free of charge or low in cost, that make many of the cartographic special effects possible. Also, thanks to ESRI founder Jack Dangermond, who took the time to give my first ArcView lesson, way back with version 1.0.

This book was produced using ArcView 3.2 on a Dell Dimension XPS T500 running Windows NT 4.0, with 256 megabytes of RAM, a 10-gigabyte ultra-SCSI hard drive, and a great eye-saving 19-inch Trinitron monitor. The text was written in Microsoft Word 97. Screen shots were taken with Jasc Paintshop Pro 5.03. Additional image and map processing was undertaken in Adobe's fine suite of graphic products: Photoshop 5.5, Illustrator 8.0, and Acrobat 4.0.

Last, a special thanks to my wife Rosemary Rowe for her endless (OK, almost endless) patience and support. Yes, I will mow the lawn and walk the dogs from now on.

Contents

INTRODUCTION . . . . **ix**

Chapter Content. ix

Features x

Companion Web Site xi

A Note to the Reader xii

CHAPTER 1
Mapping with ArcView GIS **1**

A Historical Perspective 2

What Makes a Good Map? 5

Descriptive Title 7

The Map 7

Map Legend 8

Map Scale 8

Map Projection. 9

North Arrow 10

Source Statement 10

Things to Consider Before Making a Map 10

Audience: Know Your Map Users . . 11

Know Your Data 13

Metadata. 14

Know Your Tools 15

Summary 16

CHAPTER 2
The Cartographic Design Process **17**

Symbolizing Geographic Data: The Basics 18

Vector Data 18

Raster Data 20

Labeling Geographic Data 21

Cartographic Design Tools 22

Shape 22

Size 23

Orientation 24

Texture 24

Hue 24

Value 25

Cartographic Design Principles 26

Legibility 26

Visual Contrast 27

Figure-Ground Contrast 28

Hierarchical Organization 29

The Cartographic Design Process in
ArcView 32

Map Purpose 33

List of Geographic Data Layers 33

In the View Document 34

In the View's Legend Editor 36

Back in the View 37

Sketching a Map Layout 37

In the Layout Document 38

On the Printer 38

Special Design Considerations for
Different Map Products 38

Presentation Maps and Publication
Maps .39

Thematic Maps and General
Reference Maps 40

Grayscale Versus Color Maps 41

Reflective Media and Transmissive
    Media 41
    Maps on the Web 42
Summary 43

CHAPTER 3
The Legend Editor **45**

Getting Started in the View 45
The Search for WYSIWYG: Problems
    of the 8.3 Naming Convention 48
Types of Thematic Maps 50
    Choropleth Maps 50
    Isarithmic Maps 51
    Dot Maps 51
    Proportional Symbol Maps 52
Legend Types in the Legend Editor 52
    Single Symbol Legend 55
Using Null Symbols to Show No Data . . 56
    Graduated Color 57
    Graduated Symbol 62
    Unique Value 65
    Dot Density 68
    Chart Symbol 72
Summary 73

CHAPTER 4
Classification in the Legend Editor **75**

Getting Started with Classification:
    Numerical Data Distribution 76
Classification Methods 78
    Natural Breaks Classification 79
    Quantile Classification 81
    Equal Interval Classification 82
    Equal Area Classification 84
    Standard Deviation Classification 86
Summary 87

CHAPTER 5
Palettes in the ArcView Legend Editor **89**

Introducing the Palettes 91
Cartographic Design Considerations at
    Start-up 92
    Using the Palette Manager 94
    Palette Management Options 96
Color and Fill Palettes and the
    Use of Color 97
    Colors and Fills 98
    Color Models in ArcView 100
Using the HSV, RGB, and CMYK Color
    Picker Extension 102
Calibrating Your Computer Monitor to
    Your Printer 104
Fill Palettes and the Use of Patterns 107
The Marker Palette and Point
    Symbology 109
Finding and Creating Specialized Palettes 111
The Pen Palette and Line Symbology . . . 113
Using Symbolizer to Create Line, Polygon,
    and Pattern Symbols 116
Printing ArcView Palettes with the Symbol
    Dump Script 118
Summary 120

CHAPTER 6
Typography in the View 121

Type Terminology 122
Types of Type 124
Organizing Fonts with Adobe Type
    Manager 125
Labeling Map Features — The Uses of
    Text 127
Labeling Map Features –
    Text Positioning Guidelines 130
Setting Type in ArcView 135

Contents

Creating Text for the Map Legend . 137
Labeling Map Features 141
The Auto-label Option—
    Auto-labeling One Theme 1 . . . 54
The Multi-theme Auto-label
    Extension 158
Labeling Margin Information 161
Beyond the Basics—Advanced Labeling 162
Summary 166

CHAPTER 7
Map Projections and Map Scale in the
View Document **167**

Why Do We Need Map Projections? . . 168
Projection Terminology and Concepts . 170
Projection Types 171
Spheres and Spheroids 174
Datums 176
Coordinate Systems 176
Projection Parameters 178
Choosing a Map Projection 179
Questions of Scale 180
Scale Related to Projection 182
Setting Scale in the View 184
Projections in ArcView 186
Scale and Projection Type
    Considerations 187
Projection Files and Metadata 189
Changing Projections in View
    Properties 191
Using the ArcView Projection Utility . . 200
Getting Started with the
    Projection Utility 200
Alternatives to the Projection Utility 210
Summary 212

CHAPTER 8
Cartographic Design with the
Layout Document **213**

Design Considerations 215
Page Size 215
Map Composition 215
Major and Minor Elements 216
Map Scale and Map Elements 216
Balance and Readability 218
Additional Elements and the Use of
    Neatlines 219
Organizational Style 220
Type Style for Map Elements 220
Using the Page Setup and Layout
  Properties 222
Using the Frame Tools and Properties . . 225
The View Frame 226
Legend Frame 230
Scale Bar Frame 231
North Arrow Manager 233
Chart Frame 234
Table Frame 235
Picture Frame 236
Using Graphics Tools to Organize
  Elements 238
Graphic Size and Position 238
Aligning Elements 239
Grouping, Ungrouping, and
    Simplifying 241
Neatlines 241
Drawing Graphics 242
Text Tool 243
Using the Template Manager 244
Summary 246

CHAPTER 9
Advanced Views and Layouts **247**
Advanced Techniques in the View 248
    Using Multiple Copies of the Same
        Theme 248
    Using Buffers for Cartographic Effects 250
    Generalizing and Dissolving Features 255
    Legend Tools 259
Advanced Techniques with Extensions
in the Layout 262
    Custom Legend Tool 262
    Graticules and Measured Grids
        Extension 267
    Mapper Extension for Custom
        Scale Bars and Text 270
    Saving and Restoring Documents
        Using the ODB Extension v1.2 . . 272
    View Frame Extent Nudger
        Extension 274
    Overview Extension 275
Summary 277

CHAPTER 10
Using Raster Data in the View **279**
Types of Raster Data 280
    Satellite Imagery and Aerial
        Photography 280
    Scanned Images 281
    ArcInfo GRIDs 282
Image Data Types 283
    Monochrome Images 284
    Grayscale Images 284
    Pseudocolor Images 284
    True Color Images 285
Basic ArcView Used with Raster Data . . 286

Image Legend Editor 287
    Linear Lookup Tool 288
    Interval Lookup Tool 290
    Identity Tool 290
    Image Colormap 291
Mapping with Common Image Types . .293
    Using Digital Raster Graphics 293
    Using Digital Orthoquads 296
Summary 298

CHAPTER 11
Map Output **301**
Global Considerations 302
    Color and Grayscale 302
    Page Size 303
    Margin Size 303
    Printing Time per Copy 304
    Degree of PostScript Support 307
PostScript RIPs 309
    Hardware RIPs 309
    Software RIPs 310
The Printing Process 320
    Printer Drivers 321
    Spooling and Printing 322
    Types of Printers 323
    Printing Tips 330
    Exports to Other Programs 331
ESRI Versus Adobe: Making Sense of
    PostScript Files 335
    Portable Document Format Maps . . .338
    Maps for the Web 345
    ArcView's Internet Map Server 348
Summary 350

Index 353

Introduction

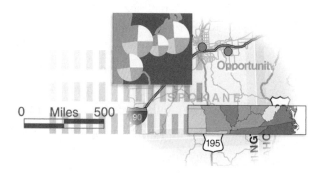

Cartographic Design Using ArcView GIS is the first book to focus solely on producing high-quality maps with this very popular desktop GIS program from the Environmental Systems Research Institute. GIS science has always been split between its analytical side and its cartographic side. It is true that analyzing geographic data is the real power of geographic information systems, but analysis is wasted if you cannot communicate results to your audience. This book is intended to help you "show what you know" in map form.

Chapter Content

This book is divided into eleven chapters. For those new to the art and science of cartography, the first two chapters deal with the basics of cartographic communication. The rest of the chapters follow in roughly the order you would follow in making a map with ArcView.

Chapters 3 and 4 explore using the powerful Legend Editor for symbolizing and classifying geographic data. Chapter 5 examines the use of the six palettes included in the Symbol Window for setting marker and line symbols, as well as the use of patterns and colors for creating area fills. Chapter 6 introduces the art of digital typography and examines the difficult task of labeling features on a map.

The complicated subject of map projections is presented in Chapter 7. Projection and scale are usually addressed at the start of the map-making process, but are placed later in the book so that those less experienced with ArcView GIS can become comfortable with the software's windows and tools before tackling this subject.

Map composition with the ArcView layout is presented in Chapter 8, and advanced design tips are examined in Chapter 9. Adding raster data, such as images, to the map is discussed in Chapter 10.

Cartographers have never had so many means of output for their map products as they do today, and Chapter 11 examines the full range of options. Laser printers, inkjets, plotters, web maps, and portable digital file formats all have different advantages, and this chapter will give the ArcView user a lot to think about.

Features

This book contains a color insert intended to present examples of the range of map output capabilities of ArcView. You will also find practical hands-on instruction within topic areas, as well as supporting features such as illustrations and diagrams, charts, tables, lists of guidelines and process steps, "sidebar" material explaining related material, and notes, tips, and warnings.

Color Insert

The sixteen plates of the color insert go beyond grayscale maps and show the full power of cartographic communication with ArcView's tools for presenting color. Plates 1 through 3 provide a quick reference to predefined color ramps, families, and palettes found in the ArcView Legend Editor.

Plates 5 through 10 display maps produced with various legend types. Plates 11 and 12 compare the visual effects of using digital raster graphics and digital aerial photographs as map backgrounds, and plates 13 through16 show advanced techniques with several of the optional ArcView extensions. In all, the color plates should give you a wealth of ideas for producing your own high-quality color maps.

Working Features

Throughout the text you will find illustrations of maps, the dialogs and other working parts of the software, and charts and diagrams—all of which are intended to inform you of where you are, what to look for, and where you can go with the result of your work. Charts and diagrams visually simplify and/or collect concepts or information otherwise difficult to perceive or reference.

Tables support the text in the same manner as charts and diagrams, and are numbered consecutively within chapters for ease of reference. Lists denote material that follows a sequential order or whose content you will want to keep track of and reference, such as process guidelines.

Notes, Tips, and Warnings work hand-in-hand with guidelines and the text in general, highlighting important points (Notes), presenting time- and hassle-saving ideas from experience (Tips), and keeping you from making mistakes that lose data or otherwise cause serious problems. "Sidebar" material, visually distinguished from the flow of text within a chapter, takes a more in-depth look at topics presented and provides information for further consideration.

Companion Web Site

Throughout the book, many free or low-cost ArcView extensions are mentioned, along with techniques on how to implement them in the map production process. You can explore these products and techniques further by going to the companion web page for *Cartographic Design Using ArcView GIS* on the Internet at the following address.

http://www.onwordpress.com/olcs/index.html

Here you will find downloadable files containing exercises based on many of the chapters in the book, as well as links to all the valuable extensions mentioned in each chapter.

A Note to the Reader

ArcView GIS has truly expanded the field of cartographic communication. It is hoped that this book will expand your skills, increase your knowledge, and save you time and headache in producing highly usable and professional-looking maps with ArcView—maps that communicate what they are supposed to, in the manner they are supposed to, and to whom they are supposed to.

It is also hoped that through this book you will have a much better understanding of the purpose of a particular map or aspect of a map in relation to your audience, and that you will be aware of, and have the skills to employ, the tools available in ArcView GIS toward these goals. With that knowledge and those skills, combined with the "practical experience" related in these pages, you will be ready for the world of cartographic design and mapping with ArcView GIS.

Chapter 1

Mapping with ArcView GIS

The technology of desktop mapping has changed radically since the introduction of ArcView 1 by the Environmental Systems Research Institute in 1992. ArcView GIS, currently in version 3.2, offers the user unprecedented ease of use for creating maps from geographic data. This coupled with the wide availability of high-quality, low-cost color printers has given mapmakers tools that could not have been dreamed of ten or twenty years ago.

Although the technology of mapmaking has dramatically changed, many of the time-tested processes of cartographic communication have not. *Cartographic Design Using ArcView GIS* introduces you to principles of cartographic design, and shows you how to apply them to mapmaking with the many tools provided by ArcView GIS.

People formally trained in geography or geographic information systems (GIS) already know of cartographic design principles, but may not have thought of how to use them with ArcView. Many in professional disciplines outside geography are starting to use ArcView to

explore their data, and need help in how to make their communication with cartographic products easier and more effective. This book
has much to offer both the experienced GIS user and the novice.

Just as purchasing a word processor for your computer does not make
you a best-selling novelist, purchasing ArcView GIS does not ensure
that you can automatically create *National Geographic* quality maps.
Cartography is a specialized form of graphic communication that
requires training, practice, and a sense of style. This book will help by
outlining the design process, which is the series of steps and decisions
the cartographer makes in the course of creating an accurate and reli-
able map. ArcView GIS has many strengths and a few weaknesses,
which this book will point out and examine as you work your way
through the cartographic design process.

A Historical Perspective

Cartography, which is the making and study of maps, predates the
invention of computerized geographic information systems (GIS) by
thousands of years. In the modern era, traditional cartographic techniques employing ink pens, plastic overlays, and press-on type were
slow and tedious, and did not allow for the generation of many alternative map products. If the map you were creating took weeks of
labor, chances are you had no extra time or energy to waste on seeing
how it would look with a different color scheme, or if the data was
symbolized using red rectangles rather than blue circles. Maps were
made by professional cartographers, who had the time and the need
to carefully plan where each map feature would be placed and to consider the visual impact of the final map product.

GIS, which organize georeferenced features from a map in a database,
became widespread in the 1980s, but still required a highly trained
professional to analyze data and produce maps from the expensive,

Chapter 1

Mapping with ArcView GIS

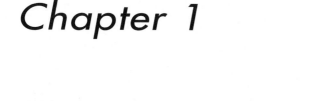

The technology of desktop mapping has changed radically since the introduction of ArcView 1 by the Environmental Systems Research Institute in 1992. ArcView GIS, currently in version 3.2, offers the user unprecedented ease of use for creating maps from geographic data. This coupled with the wide availability of high-quality, low-cost color printers has given mapmakers tools that could not have been dreamed of ten or twenty years ago.

Although the technology of mapmaking has dramatically changed, many of the time-tested processes of cartographic communication have not. *Cartographic Design Using ArcView GIS* introduces you to principles of cartographic design, and shows you how to apply them to mapmaking with the many tools provided by ArcView GIS.

People formally trained in geography or geographic information systems (GIS) already know of cartographic design principles, but may not have thought of how to use them with ArcView. Many in professional disciplines outside geography are starting to use ArcView to

explore their data, and need help in how to make their communication with cartographic products easier and more effective. This book
has much to offer both the experienced GIS user and the novice.

Just as purchasing a word processor for your computer does not make
you a best-selling novelist, purchasing ArcView GIS does not ensure
that you can automatically create *National Geographic* quality maps.
Cartography is a specialized form of graphic communication that
requires training, practice, and a sense of style. This book will help by
outlining the design process, which is the series of steps and decisions
the cartographer makes in the course of creating an accurate and reli-
able map. ArcView GIS has many strengths and a few weaknesses,
which this book will point out and examine as you work your way
through the cartographic design process.

A Historical Perspective

Cartography, which is the making and study of maps, predates the
invention of computerized geographic information systems (GIS) by
thousands of years. In the modern era, traditional cartographic techniques employing ink pens, plastic overlays, and press-on type were
slow and tedious, and did not allow for the generation of many alternative map products. If the map you were creating took weeks of
labor, chances are you had no extra time or energy to waste on seeing
how it would look with a different color scheme, or if the data was
symbolized using red rectangles rather than blue circles. Maps were
made by professional cartographers, who had the time and the need
to carefully plan where each map feature would be placed and to consider the visual impact of the final map product.

GIS, which organize georeferenced features from a map in a database,
became widespread in the 1980s, but still required a highly trained
professional to analyze data and produce maps from the expensive,

UNIX-based, computer systems. In the 1990s, technological advancements in the world of GIS made vast improvements in three major areas: computer hardware, computer software, and the availability of geographic data.

Personal computers have become faster and less expensive, allowing the user to crunch large amounts of data in minutes that would have taken hours or days if attempted ten years ago. Short-run color printing has become affordable. Putting color on maps previous to the 1990s was an expensive and daunting proposition. Even the color plotters used by GIS shops ran into the tens of thousands of dollars and required endless maintenance. Today, small printers with photorealistic color output can be found in any office supply store for a couple of hundred dollars.

GIS software has become less intimidating and less expensive. When ArcView 1 came out in 1992, it caused a sensation with its ability to visualize complex geographic data with themes that could be easily symbolized without using arcane typed-in command strings. ArcView 2 in 1994 introduced the idea of the map *layout document*, where geographic themes assembled in the view could be placed on a page and printed on a laser writer or plotter. The layout document, shown in the following illustration, streamlined the map production process from a process that took several days to just a few hours or minutes. ArcView 3.2 GIS, the current version, has improved and strengthened these features, making it the most popular desktop GIS software package today, with nearly a quarter of million copies in use.

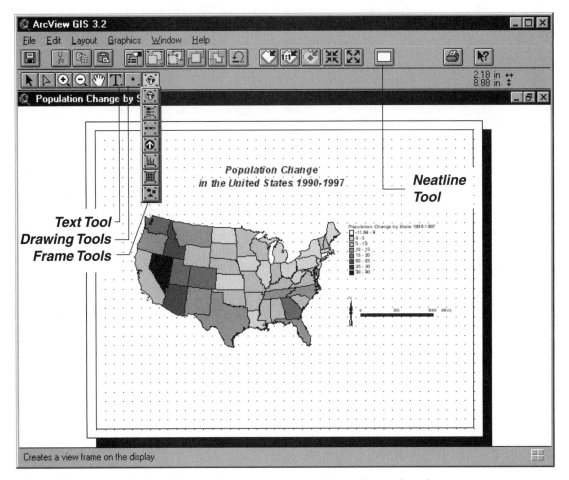

ArcView layout document with commonly used tools.

Concurrent with the development of easier-to-use GIS software has been the explosion in the availability of GIS data. Public geographic data created by government agencies free for downloading is found on hundreds of sites on the World Wide Web. CD-ROMs crammed with useful geographic data on a variety of subjects are sold by dozens of companies. Nearly all of the data can be ordered in easily used ArcView shapefile format. These trends have given thousands of people trained in disciplines other than GIS the tools to explore the geographic relationships in their data.

What Makes a Good Map?

Computer cartography with ArcView is much faster than traditional cartography and allows the generation of countless map products. You can make maps rapidly with the software, which means that you can also produce a lot of bad maps quickly. You need to know where you are going. Before examining the cartographic design process with ArcView, you need to understand what makes a good map.

Basically, there are two main categories of map: general reference maps and thematic maps. General reference maps are the types of maps you would see in an atlas, typically containing numerous map features, none of which are stressed visually over any other. Reference maps, with their rich detail, take longer to produce than other maps and require the placement of hundreds if not thousands of pieces of type.

Thematic maps are at the other end of the cartographic spectrum, and stress one or two map features over the other background items. A map showing the annual sales figures for a product by state in the United States is an example of a thematic map. The product sales figures are highlighted over any other map feature. Thematic maps are easier and faster to produce than a good general reference map, and may be the most common type of map produced with ArcView.

Maps may be classified anywhere in the spectrum between reference and thematic maps. For example, a state highway map may be rich in detail and thus resemble a reference map, but highways tend to be more prominently displayed, making it more of a thematic map. In both types of maps, certain common features make up a good map.

The following are the six elements of a good map, which are shown in the example in the illustration that follows.

- Descriptive title

- Map itself, with its symbolization of geographic features

- Legend that explains the geographic symbols

- Map scale and projection information

- North arrow or compass

- Source statement

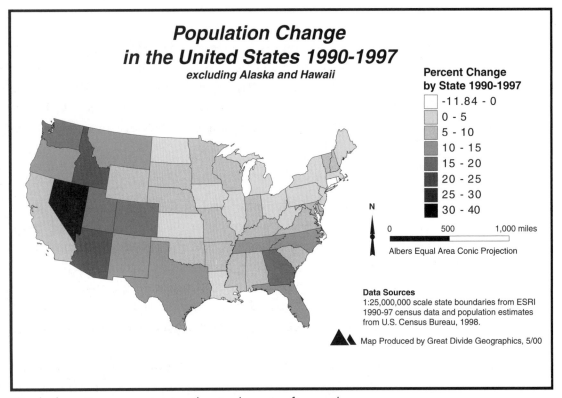

Simple thematic map containing the six elements of a good map.

These six elements form the basis of a good map. The cartographic design process provides a general roadmap that will lead the mapmaker to the final, effective map product.

Descriptive Title

A descriptive title explains in a short phrase what the map is trying to show. Consider a map of the fifty U.S. states, with each state showing estimated numeric population change between 1990 and 1999 by a graduated change in levels of gray. A quick, noninformative title for the map would be "United States Population." This title tells the map user little about the map. A better, but faulty, title for the map would be "United States Population Growth 1990-1999." Some states are losing population, and the term growth implies an increase in population, not a decrease.

Probably the best title would be "Population Change by State for the U.S., 1990-1999." This title conveys a lot of information in a short phrase and does not make the map user wonder what the map is about. In ArcView, the title is added in the layout document with the Text tool, which is examined in Chapter 8.

The Map

The map itself shows a generalization of the real-world geography in an area. Cartographers use symbols to represent real-world features, such as lines for rivers or roads, points for cities, and polygons for states or counties. Cartographers generalize geography in order to make maps clearer and easier to understand. For example, the mapmaker might choose to show only those cities with populations greater than 50,000, rather than cluttering the map with every incorporated town.

In the example of Elements of a Good Map in the previous illustration, population is generalized by state, and only the outlines of individual states (polygons) are shown. The final map is placed in the ArcView layout document using the View Frame tool, but most of the work constructing the map takes place in a view document, using the ArcView Legend Editor, shown in facing illustration.

Map Legend

A map legend clearly explains the symbols used to represent the geographic features on the map. A legend does not necessarily need to include every symbol used in the map. For example, most map readers understand that wavy blue lines represent rivers. However, the major symbols or themes you use in a map should be prominent in the legend. In ArcView there are two similar ways of constructing a legend in the layout document: using the Legend Frame tool or using the Custom Legend extension. As with the map, most of the work constructing a map legend occurs in ArcView's view document, primarily in the Table of Contents.

Map Scale

Maps always show a view of geography that is smaller than the real world, and it is necessary to note the scale of the map on the final map product. Scale can be shown as a measure, such as "1 inch to a mile"; a ratio, such as 1:24,000 scale; or as a graphic scale bar. Maps of scale 1:50,000 and less are considered large-scale maps (i.e., maps that are closer to the size of the real-world geography), whereas maps of scale 1:500,000 or greater are classed as small-scale maps. In general, large-scale maps show more geographic detail than small-scale maps. In a layout document, a scale bar is created with the Scale Frame tool.

ArcView view document and Legend Editor.

Map Projection

Map projections allow the cartographer to place a portion of the 3D curved surface of the globe on a flat, 2D piece of paper. A map projection is either set in the geographic data when it is created (and should be noted in the metadata) or set in the view if the original data is in latitude/longitude coordinates, called decimal degrees. Whatever map projection is used should be noted in a layout with the use of the Text tool.

> ➡ **NOTE:** *Map projections, which can be confusing in ArcView, are discussed in detail in Chapter 7.*

North Arrow

A North arrow or compass can be added in a layout with the use of the North Arrow Frame tool. Depending on the map's extent and projection in a view, geographic north may be directly at the top of a page or slightly to the right or left of the top. ArcView will not automatically note this in a layout, and you will frequently have to manually rotate the North arrow.

Source Statement

A source statement tells where the original map data came from and at what scale it was captured, who produced the map, and when this version was printed. This can be extensive for maps with many layers. Therefore, you might want to take advantage of the option of creating the source statement with a word processor and copying and pasting it into the layout using the Text tool.

Things to Consider Before Making a Map

Before you make a map, you should know three things: your audience for the map, your data that will construct the map, and your tools for building the map.

Audience: Know Your Map Users

ArcView can produce a wide variety of map products, from simple letter-size maps to large wall maps printed on a plotter, or map images for web pages. You should consider which map product will serve your primary audience best, and design toward that goal. Who will be the users of this map? After being in the mapmaking business awhile, you will invariably encounter a situation such as the following.

> *Client:* Can you make a simple 8.5-inch by 11-inch color map for me from the data I sent you by Friday?
>
> *Mapmaker:* Sure, no problem!
>
> *Client:* Oh, by the way, my boss would like a large version, say 3 feet by 5 feet, for a public meeting she's doing over the weekend, and a slide or overhead transparency would be great for her show when she hits the road next week, and our newsletter editor has reserved a 2-by 2-inch space for the map in the next issue. This won't cost any extra will it?
>
> *Mapmaker:* (groan) You want this map done by when?

No one map can serve all of these situations well. Some map products, such as slides or overhead transparencies, will require quite different design processes than paper maps. Everyone has been to slide presentations where a blurry image of a chart or map is up on the screen, obviously badly photographed from some paper version, while the presenter says, "If you could see this clearly, you would see this." If you cannot see it clearly, why are they showing it? The whole idea of cartographic communication is to show something clearly to your audience!

It is often good to spell out a map's purpose along with identifying its audience. Two examples are "a map of our company's annual sales figures by country for the 10,000 readers of our annual report" and "a street map of our town with points of interest for the summer tourist season." The more you can define and focus the reason for making the map, the more effective the final map product will be.

A single map can be tweaked or worked over to fit the purposes of several map products, but it takes additional time, as well as a sense of what each medium can best show. The content and visual impact of the 2- by 2-inch black-and-white map for the newsletter will be considerably different than the 3- by 5-foot full color wall map. Determine who will be the primary map users and choose a map product to best communicate your points to them. Tables 1-1 through 1-3, which follow, summarize the various map products produced by ArcView.

➤ **NOTE:** _Chapter 2 examines the various design processes in ArcView for different map products._

Table 1-1: Maps on Paper—Reflective Media

Map Sizes	English		Metric	
A Size	Letter	8.5 x 11.0 in	A4 Size	21.0 x 29.7 cm
	Legal	8.5 x 14.0 in		
B Size	Tabloid	11.0 x 17.0 in	A3 Size	29.7 x 42.0 cm
C Size		17.0 x 22.0 in	A2 Size	42.0 x 59.4 cm
D Size		22.0 x 34.0 in	A1 Size	59.4 x 84.1 cm
E Size		34.0 x 44.0 in	A0 Size	84.1 x 118.9cm
Custom Size		Whatever dimensions your printer can print		

Table 1-2: Maps on Slides—Transmissive Media

Map Sizes	English		Metric	
Camera	11.0 x 7.33 in		Matches 35 mm slide size	
A size Letter	8.5 x 11.0 in		A4 Size	21.0 x 29.7 in
	Used for overhead transparencies			

Table 1-3: Maps on Disk—Interactive and Electronic Media

File Types	Purpose
JPEG	For image maps or graphics on web pages.
PDF	Adobe Acrobat Portable Document files.
WMF	Windows Metafile for placement in other software programs.
EPS	Encapsulated PostScript file for publication quality maps and prepress work, or placement in other software programs.

Know Your Data

Ninety percent of the time invested in a typical GIS project goes to building the geographic data. When it comes time to show that data on a map, it behooves you to know the data. What projection is the data in, and what scale was it captured at? When was it gathered and who did it? What do all the cryptic names mean for the fields in the attribute table? All of these questions are answered if the geographic data comes with a good metadata record.

Metadata, which means data about data, records the important aspects of how the geographic data was produced, who the producers were, and what is contained in the data's attribute table. More data than ever is being documented according to the Federal Geographic Data Committee Standard (FGDC), a thorough national standard that any U.S. federal agency that produces geographic data is required to use.

More often than not, you will not have a complete FGDC standard metadata record to work with, because many people procrastinate documenting their data, which can be a tedious process. However, you do need to know the basics about the data set. You will also need to know the projection and scale of the data, the extent of the data (is it nationwide or just in one state or county?), and the meaning of the fields in the attribute table. It is difficult to make a map if you do not know what you are making the map with.

Metadata

As more and more people begin using geographic data in GIS systems such as ArcView, and more geographic data is being created, the importance of good metadata has increased. Metadata is the data that describes the essential elements about the geographic data. Metadata is frequently saved as a text or hypertext file that accompanies the ArcView shapefile or ArcInfo coverage that contains the real geographic data.

In recent years, many data producers have been trying to document their geographic data in accordance with the FGDC. Federal agencies producing geographic data, or an organization receiving federal funds, are required by law to document their data to this standard. This standard consists of the seven parts described in table 1-4, which follows.

Table 1-4: Recommended Sections of an FGDC Standard Metadata Document

Section	*Function*
Identification Information	Documents who produced the original data and when it was produced.
Data Quality	Indicates the scale at which the data was originally captured and the reliability of data.
Spatial Data Organization	Describes whether the data is in raster or vector format, and whether it is a series of points, lines, or polygons.
Spatial Data Reference	Records the type of projection the data is in and its coordinate system, such as an Albers projection or latitude/longitude coordinates.
Attribute Information	Defines the field names in the database and explains the words, numbers, or codes used in those fields.
Distribution Information	Indicates who is distributing the data (which can be different than who originally produced the data), and how to obtain it. For example, is it free for downloading on the web or do you have to purchase it on CD-ROM?
Metadata Reference Information	Defines what the data set should be called.

All of this information is crucial when you are trying to make a map. Without good metadata for data kept as ArcView shapefiles, you have no idea what map projection the data is in, and virtually no way of finding out. Without good metadata, you frequently will not be able to decipher the abbreviations or codes used in an attribute table, and therefore have no way of symbolizing the data correctly on a map. Without good metadata, you do not know what scale the data was captured at or in what scale it can be portrayed accurately on a map.

Luckily, today more data is being documented adequately, and the tools to do this have become much easier to use. Check out the following web pages for more information on geographic metadata, and on documentation tools.

- FGDC home page for more information on metadata standards:

  http://www.fgdc.gov/standards/standards.html

- Metadata tools home page at the University of Wisconsin:

  http://badger.state.wi.us/agencies/wlib/sco/metatool

- NOAA home page for free ArcView metadata collector extension:

  http://www.csc.noaa.gov/metadata/text/download.html

Know Your Tools

The major purpose of this book is to help you get to know your mapmaking tools, especially the GIS software. What map products can you produce with ArcView and how long will it take? ArcView is the "Swiss Army knife" of mapmaking in many respects. It does virtually anything with geographic data, but some things better than others. You have to know when the software is best employed in the map production process for certain tasks, such as adding type to a map. In some cases, it may be best to export the map to another program for finishing, or to add an ArcView extension to aid with a task. The forthcoming chapters examine the strengths and weaknesses of Arc-View, as you work through the process of designing a map.

With practice, you will become familiar with the limitations of your computer hardware and printers as they interact with ArcView. It does no good to promise a client that you can produce 100 copies of a letter-size map, including a detailed color shaded relief, in a week when you find that your super-fast laser printer takes nearly an hour to produce each copy, and when the color copier at the print shop greatly darkens all of the colors used on the map. Maps are often the most complex of graphics, and can take quite some time to print correctly, even with today's advanced printing technology.

Summary

You should now have an idea of what type of map product you want to produce, and what some of the issues are that need to be considered before the pen hits the paper. Chapter 2 introduces the design process as it applies to ArcView maps, and examines specific design problems with various map products.

Chapter 2

The Cartographic Design Process

"Good design looks right. It is simple (clear and uncomplicated). Good design is also elegant, and does not look contrived. A map should be aesthetically pleasing, thought provoking, and communicative."

—Robinson et al., *Elements of Cartography, Sixth Edition,* New York, NY, John Wiley and Sons, 1995

Producing a map that is simple, clear, and uncomplicated and that looks just right can take a lot of work. It helps to understand the basic elements of graphic design, as well as where and how to apply them in ArcView.

This chapter describes how different types of geographic data are symbolized, the representational tools that are available to symbolize data, and how those tools are applied in the cartographic design process. After describing the general cartographic design process in ArcView, design concerns for various map products are examined.

Symbolizing Geographic Data: The Basics

There are two basic ways of symbolizing geographic data, using rasters or using vectors. ArcView can use both types of geographic data, but needs different software tools to manipulate the two types.

Vector Data

With vector data (see following illustration), points, lines, or polygons represent geographic features. Years ago, traditional cartography used pens and press-on symbols to draw vector data. ArcView uses mathematical equations to draw vector features on the computer screen.

Vector data.

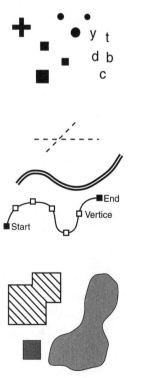

Points

Can be circles, squares, plus signs, or icons, and are dimensionless. Are measured by typographic point sizes.

Lines

Lines have starting and ending points, and vertices in between. Are measured by line weight in typographic point sizes.

Polygons

Have outlines and fills and can be simple rectangles or complex irregular closed areas.

Points can be used to represent actual physical features such as cities or towns, telephone poles, store locations, or mountain peaks. Points can also represent events (where something happens), such as auto accident sites or wildlife observation locations. Points are dimension-less locations and are marked with a wide variety of symbols, from simple circles or boxes to elaborate graphic icons. ArcView sizes point symbols by typographic point sizes, with 8 pt being the default. A size 10 pt is equal to 1/8 inch in height.

Lines typically represent linear features such as roads, rivers, power lines, or pipelines. Lines have a starting point and an ending point, as well as intermediate points. These points are called vertices, which among other things are used to represent geographic features along a line, such as stations along a subway line or the mouths of tributaries along a river. ArcView sizes line symbols by typographic point sizes, with .1 pt being the default. A tenth of a point is a very fine line that does not print legibly on many printers. Therefore, it is always good to make lines weights larger.

Polygons represent features that can be shown as areas, such as political divisions (states, counties, or zoning districts), land cover types (forested lands, grasslands), sales areas, or land ownership (public lands, land parcels). Polygons have both a fill (the color or pattern that fills the area) and an outline, measured in points.

Geographic features can be represented by different symbols when mapped at different scales. Cities, for example, would be represented by points on small-scale maps (greater than 1:500,000 scale), but might be represented by polygons to show their incorporated boundaries on large-scale maps (less than 1:100,000 scale). ArcView manipulates point sizes, line weights, and polygon fills from the various palettes found in the Legend Editor in the view document.

Raster Data

Geographic features in raster data are identified by a series of cells or pixels (picture elements) arranged in a grid with rows and columns. Each cell has three attributes: an x and y value that together identify the cell's place in the grid, and a z value, which can be either a grayscale or color value. Common examples of raster data are aerial photos, satellite images, scanned maps, and topographic surfaces. The following illustration compares raster and vector data.

Raster data compared to vector data.

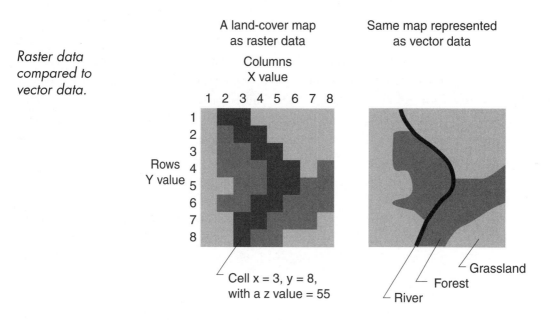

With a grayscale aerial photo, the z value will range from 0 (black) to 255 (white). Degrees of darker black tones may reflect dense evergreen forest or water, whereas lighter grays can indicate bare rock or built-up areas. In a satellite image, the z value will utilize RGB (red, green, blue) values, which are indicative of the degree of absorption or reflection of electromagnetic radiation by a feature on the ground.

A topographic surface, called a digital elevation model (DEM), has a z value that indicates elevation above sea level, in either feet or meters. Other surfaces can be created where the z value represents a cost or distance, such as distance of radio listeners from a radio transmission tower. The symbolization of the z value with either grayscale or color values can be changed in the ArcView Image Legend Editor, which differs considerably from the standard Legend Editor.

ArcView 3.1, as it comes out of the box, is essentially a vector GIS program, with minimal tools for handling raster data. Raster data can be used primarily as a background with vector data displayed on top of the image. ArcView's raster capabilities can be greatly extended with ArcView Spatial Analyst or ArcView Image Analyst, two extensions to the basic program, marketed by ESRI and ERDAS.

Labeling Geographic Data

Veteran mountain climbers know that when they have reached the summit, they are less than halfway done with their journey. Most climbing accidents happen on the descent. Likewise, experienced cartographers know that when they have finished symbolizing their data, they are only halfway done with the map. Labeling features on maps can be a tedious and demanding task. Poor or inconsistent labeling ruins many maps.

The amount of labeling of geographic features depends on the type of map. Simple thematic maps may have only a handful of named features, whereas complex reference maps may have thousands of pieces of type. With GIS data, the feature labels can be drawn from any field in a theme's attribute table. If the creator of the data has not added the labels you need, you may have to add them to the data set yourself, before making the map. In ArcView, you add text to the map and the map's legend at different points in the production process using several different tools.

➻ **NOTE:** *The use of the labeling tools and the rules governing text appearance and placement on a map are discussed in Chapter 6.*

Cartographic Design Tools

As a mapmaker, you are trying to show what you know about your geographic data. You have at your disposal a set of visual tools called visual variables that enable you to show differences in geographic data. The primary visual tools are shape, size, orientation, texture, hue (color), and value (brightness), as indicated in the facing illustration. In ArcView, nearly all of the visual variables are controlled in the Arc-View Legend Editor.

➻ **NOTE:** *The Legend Editor and visual variables are covered in detail in chapters 3, 4, and 5.*

Shape

Not only can the mapmaker show differences in geographic features by using points, lines, and polygons, but differences in shape within these three categories. Circles, squares, and triangles—or any variety of icons, such as flags, check marks, or plus signs—can represent points. Lines can be solid or dashed, single or double, smooth or jagged. Polygons can be simple rectangles or complex multi-sided areas.

Visual tools applied to geographic data.

	Points	Lines	Polygons
Shape			
Size			
Texture			
Orientation			
Hue			
Value			

Size

Points and lines can vary widely in size from small, barely visible 4-pt sizes to large 72-pt objects. Size is used to show differences in value associated with the feature. When representing cities, a large point symbol indicates a city with a larger population than a city with a small point symbol.

Polygons can range from an object representing the entire United States to an area symbolizing your back porch. Very small polygons are often best represented by points rather than areas. The mapmaker must be concerned with the legibility of the object as it appears on the final map. Will the map user be able to see it clearly? A 10-pt symbol shown as a flag may appear perfectly visible on the map when the mapmaker is staring at the computer screen from 18 inches away, but become totally lost when printed on a 3- by 4-foot map viewed on a wall from a distance of 10 feet.

Orientation

Symbols can be differentiated by their orientation to other symbols, or to the orientation of the entire map. Arrows used as point symbols can be oriented to point in the direction of the prevailing wind. Text used as labels for contour lines should be placed parallel to the slope of the hill. Arrows used with lines can be located to show direction of flow for rivers or canals.

Texture

Texture refers to the spacing and pattern of graphic symbols. The most common use of texture is with polygon fills that use dots, lines, or objects for patterns. Groupings of points in a dot density map use texture to show geographic differences, as do dashed and dotted lines.

Hue

Hue refers to the specific wavelengths of light, which we see as color (primarily values of red, green, and blue). Color can be represented by several types of color models, such as HSV (hue, saturation, and value), RGB (red, green, and blue) and CMYK (cyan, magenta, yellow, and black). HSV is ArcView's native color model, RGB is typically used by computer photo editing software, and CMYK is a standard for color printing.

     •❖ ***NOTE:*** *Color models are explained in more detail in Chapter 5.*

There are some standards in using color to represent geographic features. For land-cover maps, blue is commonly used for water, green for forest, and yellow for grasslands. With business maps, red represents retail sales and blue is used to represent manufacturing.

Color is also the most emotionally loaded of the visual variables, and needs to be used with care when considering the views or biases of your intended audience. In Rocky Mountain states such as Montana, population trends have taken divergent directions, depending on whether you live on the eastern prairies or in the western mountains.

In the case of Montana, the eastern counties of the state continue to lose population as family farms disappear, whereas the western counties experience an explosion of rural subdivisions as thousands of people move in looking for a high quality of life near Montana's trout streams and national parks. As the mapmaker producing a color map of population growth for elected officials, do you paint the counties losing population as red or the counties struggling with high growth rates as red?

Although color maps are easier to make and more common than ever before, color can be used improperly on a map and end up merely confusing the map user rather than lending understanding to the map product. For example, one proper use of color is to show differences in kind between classes of map features, not differences in amount within one class of map feature.

Value

Value determines the lightness or darkness of a color or grayscale. In general, a darker color indicates "more" of something, whereas a lighter color indicates "less." In the "Population Change in the United States" map shown in Chapter 1, states with a darker gray value have a higher numeric increase in population than states that appear whiter.

A map showing forest distribution in the northwest United States might use light green to indicate young, cut-over forest lands, and a darker green to signify dense, old growth. As an ArcView mapmaker, you will apply all of these visual tools in the design process to make an attractive and understandable map.

Cartographic Design Principles

There are four basic design principles to consider during the cartographic design process: legibility, visual contrast, figure-to-ground contrast, and hierarchical organization of layers. These principles come into play at different points in the process of constructing a map, as you are using different tools in ArcView's view and layout documents.

Legibility

If you are reading a book and have trouble understanding a paragraph, you can grab a dictionary and look up the meaning of a word or phrase, or ask another person to read the passage and get their interpretation. The author of the book may have written that paragraph poorly, but your failure to understand it does not necessarily ruin the entire book.

If you are reading a map, however, and cannot see the differences between geographic symbols or read the labels of some of the features, there is really no place to turn for help. If the mapmaker has symbolized one critical symbol poorly, the meaning of the entire map may be ruined. Maps are a unique form of graphic communication, in that their meaning hits the map reader all at once. If you have to work to understand the map's meaning, the map is worthless. This is especially true with thematic maps, but applies to general reference maps as well.

Map symbols must be legible to the reader. Lines representing roads need to be differentiated from lines representing rivers. Circular points symbolizing cities must be clearly different from points symbolizing sewage outfalls. Map feature labels should be readable by the map user under the conditions the map is designed for, whether in a book, on a wall in a large meeting room, or in a car while driving down the highway.

> ✓ **TIP:** *ArcView's Pen and Fill palettes default to a line weight of .1 pt that is very difficult to see, even if it prints clearly on your printer. Change it to at least .5 pt to be visible for lines and polygon outlines. ArcView's Font palette defaults to the first font in your type list (Arial is common on Windows computers) at 14 pt, which is moderately large for type. 10-pt type is easily readable by most people when a map is hand held by the user, whereas 6-pt is getting too small for some.*

Special considerations regarding map legibility need to be addressed for map users that may have common visual impairments, such as color blindness. Studies show that 4 percent of the American public have trouble distinguishing red from green. The one color-blind board member that cannot tell the difference on the map you made of your company's sales figures can be a problem, but producing an unreadable map of fire evacuation routes in a building can be criminal. The mapmaker uses the visual variables of size, shape, and color to help with legibility.

Visual Contrast

With thematic maps, the map symbols that represent your data should have good contrast with the other map features. The map reader's eye is drawn instantly to contrasting shapes and colors. Your job as mapmaker is to make sure the reader's eye is drawn to the features that define the purpose of the map, and is not confused with other less important information. There is less contrast between different classes of reference map features because no one feature should overpower another.

Figure-Ground Contrast

Thematic maps use visual contrast between classes of map features to establish figure-to-ground contrast. The layer or theme that contains the important data should stand out from the background layers of the map. Cartographers use several tools to trick the reader's eye into seeing the important data (the figure), which seems to "float" above the rest of the map (the ground). Figure-ground contrast is evident in the following illustration of interstate highways of the southwestern United States.

A theme used as a figure can be made darker than the rest of the map, using the visual tool of value. With size, smaller map features appear as the figure, whereas larger features constitute the ground. Map features that are closed polygons near the center of the map appear as the figure, whereas areas split by the edge of the map are discerned as ground.

Map features your audience are familiar with will tend to appear as the figure. For example, the outline of the state of Montana is readily recognized by people living in Montana, and will emerge as the figure. However, show the same map to visitors from the Far East and the state outline will disappear as the ground.

It takes some experimentation in ArcView, using a combination of Theme layering in the view and the Legend Editor to establish good figure-to-ground contrast, an example of which is shown in the following illustration. However, when you find it, you will know it. It should be easily visible!

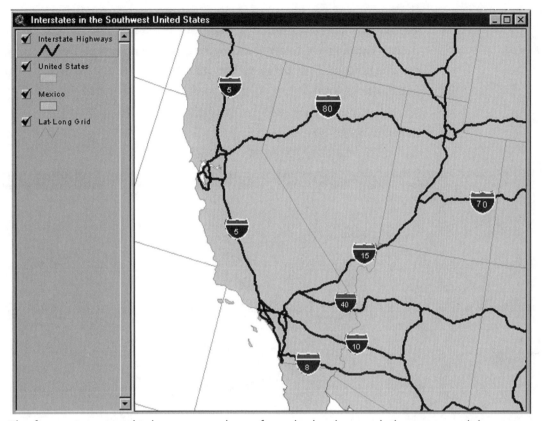

The figure, interstate highways, stands out from the background, the states and the ocean.

Hierarchical Organization

A good map is not a jumble of features but an intentionally organized series of geographic data layers. The mapmaker establishes hierarchical organization of features between themes with Theme layering in the view's Table of Contents, and within themes using the Legend Editor. There are three main types of hierarchical organization: stereogrammic, extensional, and subdivisional.

Stereogrammic organization involves organizing the layering of several themes in order to emphasize the important features of a single theme.

If done correctly, the map should appear to be organized, and data should be clearly presented at different levels within the map. Typical theme layering in an ArcView view document places polygon layers on the bottom, line layers in the middle, and point layers on the top. Establishing good figure-to-ground contrast is an example of this type of organization. The following illustration is an example of stereo-grammic organization.

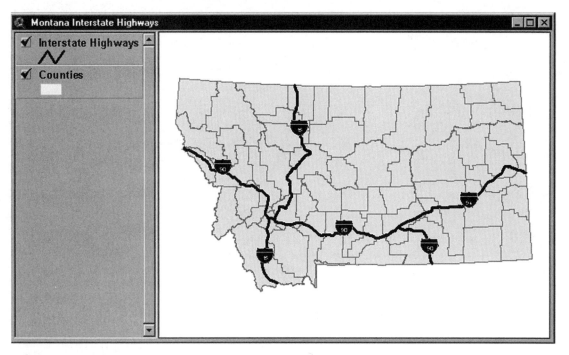

Stereogrammic organization.

Extensional organization relies on the ordering of data within a single theme, such as classifying roads with different line sizes based on whether they are local, state, or federal highways. The organization is inherent in the data; that is, it is found in a field within a theme's Attribute Table, and the mapmaker shows the organization by using differences in line or point sizes.

The organization also makes logical sense with the real-world geographic features. Local roads are shown as a smaller line weight (e.g., 0.5 pt) than federal highways (e.g., 2 pt), which reflects the reality that local roads are two lanes wide, whereas federal highways, such as interstates, may be four or more lanes wide. The following illustration is an example of extensional organization.

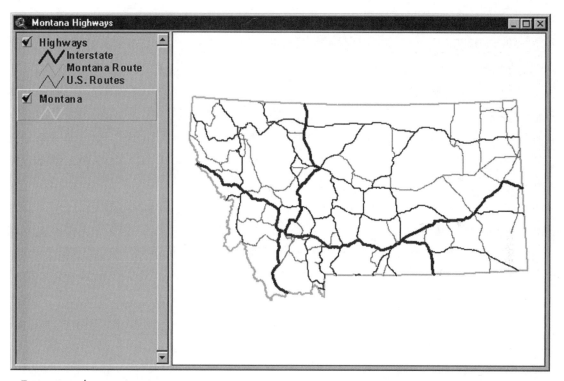

Extensional organization.

Subdivisional organization is similar to extensional organization, but applies primarily to polygon themes. Land-cover mapping provides an example here: polygons representing types of evergreen forests can be shown in various shades of dark green, whereas grasslands might be portrayed in yellows. The map reader instantly recognizes the differences between forest and grasslands—something they could not do easily if the different types of evergreen forest were shown as greens, reds, and browns, and grasslands shown as yellows and blues. The following illustration is an example of subdivisional organization.

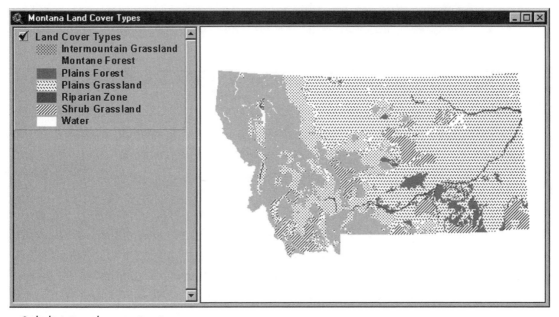

Subdivisional organization.

It is up to the mapmaker to know the data he or she is trying to represent, and to organize it logically for the map reader.

The Cartographic Design Process in ArcView

With cartographic design principles in mind, and a knowledge of the basic tools of mapmaking, you are ready to design a map in ArcView. The following section works through a general design process for ArcView map production, and the balance of the book examines the individual stages of the process in much greater detail, on a chapter-by-chapter basis.

Map Purpose

Before even starting the ArcView program, you have determined the map's purpose and have chosen an appropriate map product to produce for your intended audience. A good exercise for determining the map's purpose is writing down the possible phrases that can be used as the descriptive title of the map, such as "1998 Annual Sales for the ABC Company in the Southeast United States" or "Large-scale Vegetation Map of the Mono Lake District in California." You know whether your final map product will be a thematic map, a general reference map, a presentation map, or a publication map.

List of Geographic Data Layers

Now it is time to make a list on paper of the spatial data layers you need to accomplish your map's purpose, which are the layers you will represent as themes in the view document. If you are lucky, or perhaps just better organized than most, you might have all of the data you need already available on your computer's hard drive or somewhere on your office's network. If not, you should check the easily available data sources first, such as the CD-ROMs of data included with the ArcView program package.

> ✓ **TIP:** *Going to the Internet, check out the data section of the ESRI web site at* http://www.esri.com/data/index.html *or consult a hotlist of state geographic data centers at* http://nris.state.mt.us/gis/other.html.

Commonly, you may have to construct custom spatial data, by digitizing a paper map into the computer, digitizing special features on screen within ArcView, or bringing tabular data from an external database into ArcView and linking or joining the data to spatial features.

↠ **NOTE:** *See* INSIDE ArcView GIS *(OnWord Press) for more information on extending data sources for ArcView.*

Organize the geographic data on your computer system so that it is easily accessible to ArcView, either locally or across your office network. Remember, 80 percent of the work involved in a GIS project can be finding or creating the data you need before you begin producing maps!

In the View Document

Once your data is organized on your computer, you can begin adding data as themes in the view, using the Add Theme button. Now is the time to try some initial layering options, making themes visible and moving different themes onto different levels in the Table of Contents in order to achieve hierarchical organization of map features. Do not spend too much time layering themes at this point, before the data is properly symbolized with the Legend Editor.

Next, zoom in and out of the view window, until you reach a map extent or map scale that represents the area of the Earth you want to cover in your final map. The distorting effects of differing map projections need to be considered as well. Map projections can be changed within the standard ArcView package for certain types of data, or re-projections can be accomplished with other software packages. Table 2-1, which follows, provides checklists of the ArcView design process.

Table 2-1: ArcView Cartographic Design Process Checklist - Part 1

Location and Tool	Task
On paper:	Determine map purpose and audience
	Choose appropriate map product
	List data needed to accomplish map purpose
On your hard drive:	Gather and organize data

Location and Tool	Task
In the AV view document:	
Table of Contents and Map Display	Add data to a view
	Arrange theme layering order
Theme Properties	Label theme names
Map Display	Determine map scale and view extent
View Properties	Determine map projection
Legend Editor	Symbolize data in themes
	• Choose a Legend Type
	• Choose a classification method
	• Change symbology and color
	• Add text to a Legend
Table of Contents and Map Display	Adjust theme layering for figure-ground contrast
TOC Style in View Menu	Apply type styles to Table of Contents
Label Tool	Label selected map features in the view

Table 2-1: ArcView Cartographic Design Process Checklist - Part 2

Location and Tool	Task
On paper:	Sketch draft map layout
In the AV layout document:	Create layout document
• Frame Tools	• Place frames for map elements
• View Frame	• View
• Scale Frame	• Scale
• North Arrow Frame	• North arrow
• Text Tool	• Map title
• Legend Frame or Legend Extension	• Legend
• Text Tool	• Source information
• Neatline Button	• Add neatlines

Table 2-1: ArcView Cartographic Design Process Checklist - Part 2

Location and Tool	Task
On printer:	Print draft on printer
In the AV layout document:	Revise layout and print final map

In the View's Legend Editor

This is the heart of the matter, where you take the time to symbolize your data in ArcView's powerful Legend Editor. Here, you have to know what all those column names mean in the theme's attribute table. The Legend Editor can slice, dice, and symbolize geographic data in literally countless ways, so it is crucial to understand what type of data you are working with. Is it made up of points, lines, or polygons? Is the data categorical or numerical? Are attribute fields using ordinal, nominal, interval, or ratio levels of measurement?

⇝ **NOTE:** *Chapters 3, 4, and 5 thoroughly examine the use of the Legend Editor.*

If you understand your data very well, know what you want to show on the map, and are familiar with the tools of the Legend Editor, this step can go quickly. However, do not hesitate to take the time to explore your data's attributes and test various ways of visualizing your geographic data.

First, choose a Legend Type from among single symbol, unique value, graduated color, graduated symbol, dot density, or chart symbol types. Next, decide on a classification method, if applicable, and determine the number of classes you want to appear in the theme's legend. Experiment with changing symbology and color to enhance the data's visual contrast. Finally, add text to the legend to clearly explain the data classes you have chosen.

Back in the View

Returning to the view, adjust Theme layering to fine tune data visualization. With thematic maps, apply the principle of figure-to-ground contrast to cause the important layers to stand out from the map background. Alternate between theme layering in the Table of Contents and data symbolization in the Legend Editor to achieve the effect you want in the view's Map Display. Remember, the way your data appears in the view's Map Display is the way the map will appear in ArcView's layout document.

After the data themes are symbolized in the manner that meets your map's purpose (and does not stretch the data's validity!), begin adding text to the Map Display by labeling features in the view with the Label tools. Typography on the map can take a considerable amount of time, depending on the type of map and the map's purpose. When you are finished here, you are ready to proceed with laying out the final map.

⟜ **NOTE:** *Chapter 6 explains how to establish a hierarchy of labels to make maps easily readable.*

Sketching a Map Layout

Review the elements that make up a good map. On a piece of paper, sketch where you would place each map element. Look at other printed maps similar to the one you are constructing, and think of how effectively or ineffectively they make their point. Consider the final size of your map product. Do you have room for a large legend to the right of the map frame, in a 4- by 4-inch map? Can the map title fit in the 4-inch square box or will it have to appear as a caption when the map is placed on a letter-size page in an annual report?

In the Layout Document

Open a new layout document in your ArcView project, and choose a page size for your final map product. Although you are selecting a page size in ArcView at this point for your map product, you should have decided upon a final size of the map product at the start of the design process, when you determined the map's purpose and audience. Otherwise, you can end up with far too much detail to be legible when that tiny map is finally printed!

Following your draft sketch of the map layout, begin placing frames with the Frame tool in the layout for the map, the map legend, the North arrow, and the scale bar. Add text for the map title, source information, and projection data. Consider adding neatlines and graphics to enhance the map layout.

On the Printer

Print a first draft of your map product. Does it show what you intended to show? Are the grayscales or colors too intense or misleading? Are all of the significant map features labeled correctly and legibly? Show the draft map to a colleague or the client and ask them whether it achieves your intended purpose. Go back to the layout or view and change the elements that need to be shown more clearly. Print your revised map product.

Special Design Considerations for Different Map Products

The general design process in ArcView needs to be tweaked for different types of maps. In some cases, you might need to export the layout and put the finishing touches on the map in another software program, such as Adobe Illustrator or MacroMedia Freehand. The following sections discuss design considerations to keep in mind while preparing different types of maps.

Presentation Maps and Publication Maps

ArcView excels at producing presentation-quality maps; that is, maps printed on personal lasers or inkjets, and included in a report or document or perhaps displayed at a meeting. Presentation maps are typically short-run print jobs, with anywhere from 1 to 500 copies produced for distribution. The general design process previously outlined works well for presentation-quality maps.

Once the decision is made that more than about 500 color copies of the map are needed, it becomes reasonable to go to a printer to set up and publish your map on a printing press. If your map product is destined to be published in a magazine, a book, or an atlas, you will need to produce a publication-quality map. Publication-quality maps require more care in using type and color, and in preparing the final computer file for printing.

A print shop will first print your map on a device called a digital image setter, and create high-resolution color separations on film based on the number of colors used in the map. Then, the film separations are used to generate plates, which are placed on the printing press to trap different inks in the printing process. Some image setters skip the film step and generate plates directly. Nearly all image setters require Encapsulated PostScript files (EPS files) to generate film correctly. ArcView offers a PostScript Export option in the File menu within the layout document.

Before reaching the file export point in the layout, additional care should be taken in placing type on the map and assigning color. If the map requires a lot of type (several hundred labels), it is frequently best to export the map from the ArcView layout without any labels and place the type in a PostScript graphics program.

> ✦ **NOTE:** *See Chapter 6 for more information on typography in Arc-View.*

Colors are assigned in ArcView based on a hue, saturation, and value (HSV) color model. Colors are commonly handled in the publication printing process as a cyan, magenta, yellow, and black (CMYK) color model. ArcView's color palettes can be changed to a CMYK model, or colors can be assigned in a PostScript graphics program after the map is exported as an EPS file from the layout document.

> ✦ **NOTE:** *See Chapter 5 for more information on assigning color and color models in ArcView.*

Thematic Maps and General Reference Maps

In general, thematic maps are less complex than general reference maps and have fewer map features and labels. Thematic maps require more visual contrast than general reference maps, and their message is meant to be conveyed as a whole and all at once to the map reader. A map user may study a detailed highway map (primarily a general reference map) to find a route through a city, but if the map reader does not quickly understand the point of your company's 2000 annual sales performance map (a thematic map), the map is useless.

Thematic maps need to go through the entire design process in Arc-View before a draft can be generated and the map product examined to see if the map's purpose is being met. With a detailed general reference map, it is better to produce a small portion of the map as a draft and examine the placement, size, and color of features for legibility. You want to be able to find out that your highway symbols are too small to be readable on a draft 4- by 4-inch portion of the state of California before placing thousands of them for the entire state.

Grayscale Versus Color Maps

The human eye can detect anywhere from three to seven different grayscale values on a map, depending on the resolution of the printer generating the map. The eye can detect literally thousands of color values, but using that many colors can be quite confusing. Often, simpler is better!

Thematic maps are based on the visual contrast between map features. You need to know whether your printing device will show the same visual contrast on the paper map as ArcView shows on the map in the view document. Before assigning grayscale values or color values to map features, try printing out a grayscale or color chart from ArcView (discussed in Chapter 5) to determine what values will produce the contrast you need on your printer.

Reflective Media and Transmissive Media

Paper maps reflect light from the surface of the paper, depending on the color of the ink used. Darker colors are used to make map features stand out from lightly colored background features.

Slides and overhead transparencies use light transmitted through the media they are printed on. They are also similar to images shown from a projector connected to a computer. Lightly colored features stand out more than those with darker colors. In fact, dark lines less than 4 pt in size and dark text smaller than 14 pt can be very difficult to read. This is why presentations using slides or transparencies hastily produced from paper maps are so difficult to see. Practice the color reversal technique for transmissive media; reverse the color schemes used on paper maps for slide presentations, and make features stand out with lighter color values.

> ↦ **NOTE:** *Chapter 5 discusses ArcView palettes for specific use with transmissive media.*

Slides, overheads, and projected maps also have much less "linger time" for the viewer than paper maps. That is, the map reader does not have much time to examine a complex legend or read dozens of labels for map features before the presenter has moved on to the next image. Transmissive maps need to be much simpler than their paper counterparts, with less type and fewer map features. This is a good point to stress the next time a client or boss asks for a quick slide from an existing paper map. You need the time to communicate effectively using the strengths of each type of media.

Maps on the Web

For literally hundreds of years, maps have been produced only on paper. The World Wide Web of the 1990s represents a new medium available to mapmakers. Maps on the Web can be static or interactive. Static maps are typically GIF or JPEG images of maps, embedded into web pages that appear in your web browser. The general cartographic design process in ArcView, tempered with considerations for transmissive media, serves well for static web maps.

Maps produced for paper products can be printed from the ArcView layout to Adobe Acrobat Portable Document Format (PDF files) and served on the Internet as static maps.

> ↦ **NOTE:** *Chapter 11 discusses generating PDF files from ArcView layouts in more detail.*

Interactive maps can appear as image maps on a web page. Image maps have clickable parts embedded in the map, which take the map viewer to another web page when clicked on with a mouse.

Interactive maps can also be database driven, operating almost identically to the manner in which map features in views and tables interact in ArcView. The ArcView Internet Map Server Extension (AVIMS) allows the mapmaker to serve ArcView views directly on the Internet.

> **NOTE:** *Chapter 9 examines some design considerations for image maps and AVIMS maps.*

Summary

As a cartographer, you have many visual design tools at your disposal for correctly presenting geographic data. It is good to be grounded in use of basic graphic design elements in order to understand their proper use before applying them with the powerful tools supplied by ArcView. The next several chapters will go into detail on how to use the ArcView tools in the cartographic design process.

Chapter 3

The Legend Editor

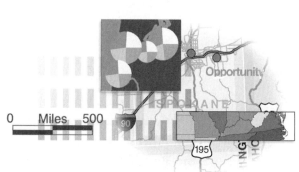

This chapter and chapters 4 and 5 focus on the powerful Legend Editor in ArcView's view document. The Legend Editor can be used to explore the relationships within your geographic data, and is used to symbolize the data as it will appear on the final map. It is truly the "Vegomatic" tool of ArcView, capable of slicing and dicing your data in nearly endless ways.

Getting Started in the View

The Legend Editor, in concert with several other tools in the view document, is responsible for setting up the map display the way you want it to appear on the final map product. Before diving into the intricacies of the Legend Editor, examine a couple of the other tools in the view.

You have already determined the purpose and audience of your map, and have organized the appropriate geographic data on the computer's hard drive or another hard drive accessible over the local network. The geographic data can be in several standard formats: ArcView shapefiles, ArcInfo coverages, event themes created from tabular data, CAD files, and many types of image file formats.

➥ **NOTE:** *See Chapter 10 for more detail on using imagery in the view. If your data takes the form of more than one map projection, see Chapter 7, on how to change map projections.*

After an ArcView project has been started, and a new view opened, you can begin adding data to the view with the Add Theme button. The Add Theme dialog box opens, which allows you to navigate the computer's directory system on the right side of the box, and shows the available geographic data suitable for making a theme on the left side of the box. The Add Theme dialog box is shown in the following illustration.

Add Theme dialog box.

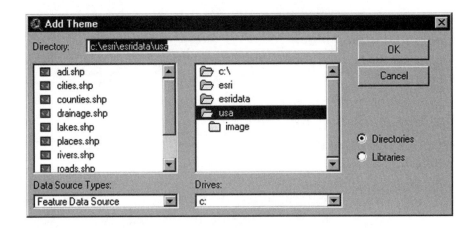

ArcView adds new themes one on top of each other in the Table of Contents, with the bottom-most theme shown first and the topmost theme shown last. You can shuffle theme order by clicking and dragging the theme's name and placing the theme further up or down in the Table of Contents. In general, you want filled polygon themes on the bottom of the stack, point themes on the top, and line themes in the middle, as shown in the following illustration.

A simple Table of Contents with polygon themes on the bottom, line themes in the middle, and point themes on the top.

☑ **cities.shp**
•
☑ **roads.shp**
∿
☑ **states50.shp**
▭

There are times when you may want to place the same theme several times in a view. For example, with political subdivisions such as states and counties, you may want to have a filled polygon of counties on the bottom of the TOC, and an unfilled polygon of the same theme on the top (with the polygon's outline emphasized). In addition, you may symbolize a single theme in multiple ways using different theme attributes in the Legend Editor.

✗ WARNING: *Know when you are editing a theme that has been copied several times in a view using Copy and Paste from the Edit menu. This will kill your project and destroy the theme's original data!*

The previous warning concerns what is perhaps the most common cause of corrupting an ArcView project and destroying data. This happens because ArcView cannot figure out how to edit a theme while still displaying its pre-edited condition from the same file. Play it safe and always use the Convert to Shapefile command in the Theme menu while in the view to make a copy of the original data, and then edit only the copied shapefile. It is also good practice to place the copies of themes you want to edit in a special edit directory on your hard drive, to avoid changing an original data set inadvertently.

Copy and paste in the Edit menu. Do not create a new copy of the original data in a theme on your hard drive, but create only a new link to the original data set. Remember, an ArcView Project file (an *.apr* file) contains no geographic data. An ArcView Project file contains pointers to the geographic data on your hard drive or network, and instructions on how to symbolize and display that data.

The Search for WYSIWYG:
Problems of the 8.3 Naming Convention

Back in the dark ages of personal computing, not that long ago, computers were slow and RAM memory was an expensive commodity. Following their mainframe computing ancestors, who had used a shortcut to record an important date in order to minimize memory requirements, programmers of personal computing software limited file names to the 13-character naming convention, the so-called "8 dot, 3 file" name. Computer users were forever attempting to think of creative, meaningful abbreviations for their computer files, while being strait-jacketed by the 8.3-character limitation.

In a parallel universe of computing, namely that of the Macintosh computers, programmers were pursuing the holy grail of graphic computing: WYSIWYG. This acronym stands for "What you see is what you get." This radical notion posited that what appears on the computer's screen should be what appears on paper when you print the document.

These two diametrically opposed philosophies of computing clash in the view's Table of Contents. Geographic data, especially ArcView shapefiles, are still named according to the 8.3 naming convention, although Windows 95/98 and NT allow file names to be longer. Shapefiles must have the proper three-character suffixes *(.shp, .shx, and .dbf)*, all three of which must be found in the same directory. Naming the first part of the name longer than eight characters may cause the shapefile to be unusable in certain ArcView processes. Likewise, exporting a *.dbf* table with a first name longer than eight characters will make the table unusable in many Microsoft products, such as Microsoft Access.

For naming the actual shapefile on your hard drive, use only alphanumeric characters, dashes, and underlines in the 8.3 name. Using anything else will make the shapefile data unusable. In addition, if you plan on using the shapefile in ArcInfo, the ArcInfo program does not like numbers at the start of a file name.

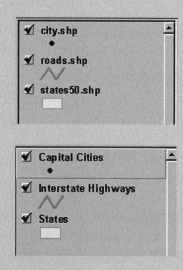

View Table of Contents before and after renaming themes.

The view's Table of Contents attempts to fulfill the goal of WYSIWYG, with the TOC being reflected "as is" in the final map legend placed in the ArcView layout. As a result, if you do not rename your themes in the view, you end up with incomprehensible legends on your final map, with notations such as *usa90pop.shp* and *lwcntyrd.shp.* The illustrations at left show the view Table of Contents before and after themes are renamed.

As you begin to add themes to a view, you should also begin to rename the themes in the view using the Theme Properties dialog box in the Theme menu. Thus, *usa90pop.shp* becomes "United States Population – 1990," and *lwcntyrd.shp* becomes "Lewis and Clark County Roads." This will not change the name of the original data file on your hard drive. However, it will make your map much more understandable. Follow this simple rule when constructing a view: If another person looking over your shoulder cannot quickly understand what the data are that you have placed in the view by reading the theme names in the Table of Contents, you are not making a good map.

Types of Thematic Maps

The following are the four basic types of thematic maps. These can all be created in ArcView using the Legend Editor.

- Choropleth maps

- Isarithmic maps

- Dot density maps

- Proportional symbol maps

Choropleth Maps

Choropleth maps are used to show changes in numeric data values by area (polygon), by shading the area with different intensities of gray or a single color. Different intensities of gray can be used to show human population growth by state, for example. Growth of a single agricultural crop, such as wheat, can be represented by a scale of green, from light green representing little growth, to dark green, showing areas with large amounts of wheat production. ArcView uses color ramps in the Graduated Color legend type to create a choropleth map.

Choropleth maps can be misleading, unless the numeric data is normalized in some fashion. To normalize data, the raw data you want to represent can be related to a second numeric attribute. Wheat production by state, for example, can be normalized to the size of the state, or population growth can be normalized to the size of the area, resulting in population density. In the ArcView Legend Editor, raw numeric data can be normalized by relating the attribute to another numeric attribute, or by presenting the data as a percentage of the total.

Isarithmic Maps

Isarithmic maps show a series of lines, called isolines, drawn between sample points of equal value. Isarithmic maps are commonly called contour maps. A topographic map shows contours that connect points on the earth's surface with the same elevation above sea level, measured in meters or feet.

Elevation is not the only data that can be used to create a surface with contours. Nearly any numeric data, sampled at a fine enough interval, can be used to create a surface. Annual rainfall data can be mapped to show precipitation, where each contour represents an additional inch (2.54 centimeters) of rainfall.

Basic ArcView can symbolize contour lines in the Legend Editor, but it cannot create them from raw data. In order to generate a surface from raw data, you need the tools provided by ArcView Spatial Analyst, an optional extension produced by ESRI.

A similar form of isarithmic map could show distances as contours from a factory generating a form of air pollution. These equal distance contours can be created in basic ArcView using the Create Buffers command in the Theme menu, and then symbolized in the Legend Editor.

Dot Maps

With dot mapping, dots (actually small, solid circles) of equal size are used to represent an amount of a certain attribute, and are then placed near where that attribute may be found. This form of dot mapping is normally used with area (polygon) data, and can be more informative than a choropleth map of the same attribute. In ArcView, the map-maker assigns the number of the attribute data per dot in the Legend Editor, and the software uses a random number generator to produce a dot density map.

Proportional Symbol Maps

Proportional symbol maps use raw count data from an attribute field to create a symbol sized according to the value of the attribute field. The symbol is then used to represent that value for an area (in the center of a polygon), or the value is used to scale a line or a point feature. The ArcView Legend Editor allows the user to create two types of proportional symbol maps: Graduated Symbol legends for points and lines, and Chart Maps, which are best for polygons, but can be used for points and lines.

Legend Types in the Legend Editor

The Legend Editor is accessed one of three ways: by double clicking on a theme's name in the TOC, clicking once on the Edit Legend button on the view button bar, or by selecting Edit Legend from the Theme menu. The basic Legend Editor appears, showing six buttons initially. The facing illustration shows the three ways of accessing the Legend Editor.

Using the Legend Editor is a four-step process.

1. Select a legend type (explained in the following paragraph).

2. Select a classification method (explained in Chapter 4).

3. Change symbology and color (explained in Chapter 5).

4. Add descriptive text to the legend (explained in Chapter 6).

*Three ways of accessing
the Legend Editor.*

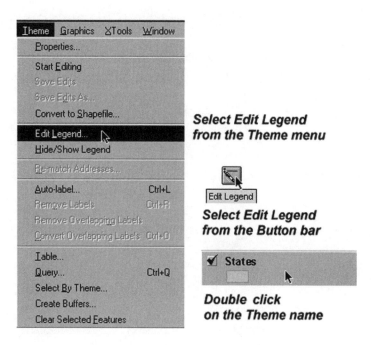

**Select Edit Legend
from the Theme menu**

**Select Edit Legend
from the Button bar**

**Double click
on the Theme name**

Once the first step is taken and a legend type is chosen, more options become available within the Legend Editor, shown in the following illustration. ArcView offers six basic Legend Types: single symbol, graduated color, graduated symbol, unique value, dot density, and chart. Only five Legend Types appear at any one time within the Legend Editor, because some types, such as dot, are only available with polygons, whereas others (graduated symbol) are only available for point and line themes. Table 3-1, which follows, lists and describes the ArcView legend types.

Basic Legend Editor showing a single symbol legend type, and the Legend Type drop-down list.

Table 3-1: ArcView Legend Types

Legend Type	Points	Lines	Polygons	Notes
Single Symbol	Yes	Yes	Yes	ArcView default
Unique Value	Yes	Yes	Yes	Good for nominal data
Graduated Color	Yes	Yes	Yes	For numeric data only; best with polygons
Graduated Symbol	Yes	Yes	No[1]	For numeric data only
Dot Density	No	No	Yes	Only with polygons
Chart Symbol	Yes	Yes	Yes	Best with polygons

[1] The Graduated Symbol legend type can be used with polygons if polygon centroids are created first.

Single Symbol Legend

The Single Symbol legend type is ArcView's default legend type, and symbolizes all features within a theme as one color, and one point size (for point and line themes, and polygon outlines). This is useful for showing the basic geographic location of the data within a theme, without breaking down the data in a more complicated manner. In most mapping situations, you will want to move on to another legend type in order to symbolize the data in a more meaningful way.

However, Single Symbol is appropriate for background or base map themes, where you are trying to establish a "ground" for figure-to-ground contrast (see Chapter 2). Typical base layers where Single Symbol legends would be applied are any sort of political boundaries, such as countries, states, provinces, or counties. The more important geographic data in thematic maps (the figure) should be symbolized with one of the other legend types.

Single Symbol is also useful with polygon themes symbolized using only the polygon outlines for reference. Frequently, an important theme will be symbolized using the graduated color legend type, whereas another polygon theme containing the outlines of political subdivisions will be placed on top of the initial theme.

You can change the polygon fill and outline by double clicking on the symbol in the Legend Editor (see Chapter 5) or by single clicking on the Default button, which will cycle through a number of colors.

Using Null Symbols to Show No Data

The main purpose of mapping with ArcView is to show what you know about your data. However, in many cases, it is also important to show what you do not know. To do this, the ArcView Legend Editor provides a way of setting "null value" symbols.

Null values are essentially any values you chose not to include in your symbolization while using the Legend Editor. For character string fields, a record with a null value may be blank, or have the words *none* or *not sampled* typed in. If the field is blank, the ArcView Legend Editor will automatically ignore the field and a null value will not have to be set. With numeric fields, common null values are 0, –99, or –999.

ArcView allows the user to set any number of null value measures for a field in the Legend Editor by setting a null value in the Null Value dialog box, and choosing whether to display the null value in the legend. Null values may reflect any number of situations, such as a lake or well that has not been sampled and measured for a particular water quality parameter, a store outlet that has not reported its sales figures, or a polling place that has not tabulated its votes. In environmental situations, a rare plant may be found in one half of an area, whereas the other half may not have been surveyed as yet.

With these examples, it is important to note that the values are not zero, which would reflect that something has been measured, and found to be 0. Rather, no value has been measured as yet at these locations. Frequently, it is equally important to show on a map where you have not measured something as it is to show on the map where you have measured something. A popular example found on weather maps on television or the World Wide Web is the use of advanced Doppler radar to indicate rainfall intensity across the United States. Doppler radar sites are distributed across the United States in a pattern designed to overlap the radar coverages and avoid having major blank areas between sites.

Few weather maps, however, will ever indicate where those blank areas may be. For example, according to Doppler radar for Washington state, it never rains on much of the Olympic Peninsula west of Puget Sound. In reality, the western side of the Olympic Peninsula is the wettest region of the lower 48 states, but is in the "radar shadow" of the rugged Olympic Mountains, which block detection of rainfall events from the Doppler radar at the Seattle-Tacoma International Airport to the east. When choosing what null values or no data values to show or not to show on a map, care should be taken not to mislead the map user.

Graduated Color

The Graduated Color legend type works best with polygon themes for showing changes in magnitude for a particular theme attribute. Graduated color also works with point and line themes, but the small size of many point and line symbols may make the changes in color difficult for the map user to interpret.

Once Graduated Color is chosen in the Legend Editor's Legend Type drop-down list, shown in the following illustration, two other options become available. The Classification Field drop-down list appears, and a field from the theme's attribute table must be chosen to apply the Graduated Color ramp to. Directly below this drop-down list is the Normalize option.

> ✓ **TIP:** *Graduated Color legends should always be normalized, by relating the attribute to another numeric attribute or by presenting the data as a percentage of the total.*

Legend Editor and the Graduated Color legend type.

A common example where normalization should be applied is population data. Raw population counts can be symbolized using Graduated Color ramps for U.S. states, which show that California has the highest population, and Wyoming the smallest. However, no two states have the same area, so the raw population count data can be normalized against the state's area, giving a much more useful legend showing population density throughout the United States. Raw count data can also be normalized as Percent of Total, which will calculate the total number within a numeric attribute field, and then show each class's percentage of that total.

The Legend Editor defaults to five classes, the number of which can be changed by adding or deleting classes using the Add Class or

Delete Class buttons, and typing new values into the Legend Editor's Value column.

➥ **NOTE:** *Chapter 4 examines classification strategies in detail.*

Clicking on the Values or Labels (which can switch to Counts for certain types of attributes) column headings allows you to apply the Sort Descending or Sort Ascending buttons near the bottom of the Legend Editor. The Flip Symbols button will switch the order of the Symbols for each class, but will leave the Values and Labels orders unchanged.

The Graduated Color Legend Editor makes available a wide range of predefined color ramps, which are accessed from the Color Ramp drop-down list. You can use these predefined color ramps or enter your own, using the Ramp Colors button. The predefined color ramps are of four basic types: single-color monochromatic, two-color monochromatic, two-color, and specialized geographic. The following is a list of the predefined graduated color ramps in the Legend Editor.

- Single-color Monochromatic Color Ramps

    - Red monochromatic
    - Orange monochromatic
    - Yellow monochromatic
    - Chartreuse monochromatic
    - Green monochromatic
    - Blue-green monochromatic
    - Cyan monochromatic
    - Blue monochromatic
    - Magenta monochromatic
    - Gray monochromatic

- Two-color Dichromatic Color Ramps

  - Blues to Reds dichromatic
  - Grays to Reds dichromatic
  - Blues to Oranges dichromatic
  - Purples to Oranges dichromatic
  - Greens to Reds dichromatic
  - Greens to Oranges dichromatic
  - Browns to Blue-greens dichromatic
  - Purples to Greens dichromatic
  - Cyans to Reds dichromatic
  - Cyans to Oranges dichromatic

- Two- and Three-color Ramps

  - Yellow to Orange to Red
  - Beige to Brown
  - Red to Purple to Blue
  - Green to Cyan to Blue
  - Yellow to Green to Dark Blue

- Specialized Geographic Color Ramps

  - Elevations #1
  - Elevations #2
  - Sea Floor Elevation
  - Full Spectrum
  - Precipitation
  - Temperature
  - Land Cover #1
  - Land Cover #2

The single-color monochromatic ramps symbols within one color family—from light red to dark red, for example. This is useful for showing numeric data that increases from a zero point to some larger number. Crop production, such as bushels of wheat or corn per acre (or hectare), works well with a monochromatic ramp. Low numbers or zero indicates crop failure (shown as a lighter, whiter color), whereas high numbers shown in intense, saturated colors reveal a bumper crop.

The two-color dichromatic ramp allows you to represent bivariate data distributions, by going from one color to white, then up to another color. Population growth rates of U.S. states or counties present a good example of a bivariate data distribution. Population is not increasing everywhere in the United States. Some states and counties are losing population, others are gaining, and some exhibit little change at all over the last decade. The blues-to-reds dichromatic ramp can be used to show these trends, with blues representing areas with population losses, the white class revealing areas with little or no change in population over time, and the more intense reds representing areas of high growth.

The two-color ramps are similar to the two-color dichromatic ramps, but do not pass through a class shown as white in the middle of the distribution. This makes them somewhat less useful in representing bivariate data.

There are seven ramps for specialized types of geography: Elevation 1 and 2, Sea Floor Elevation, Precipitation, Temperature, and Land Cover 1 and 2. These ramps provide standard color schemes for geographic phenomena, such as hypsometric tints for terrestrial elevations, ranging from greens at low elevation to whites on mountain peaks; browns, reds, and blues for precipitation; and shades of greens and browns for land cover classifications.

These ramps use different colors as applied to 13 classes. Therefore, it is helpful to split your data into 13 classes using the Classify button before applying the specialized geographic ramps. Last, a Full Spectrum color ramp is provided, which will attempt to represent a wide range of colors scattered across the visible light spectrum.

The Ramp Colors button allows you to quickly create your own customized color ramps. You can choose a color for your first class in the Legend Editor, choose a color for your last class (red and blue, for example), and click on the Ramp Colors button and ArcView will assign colors that fall between red and blue on the color spectrum for the intervening classes. You can also assign a color for a class in the middle of the Legend Editor's tabular display, select that class by clicking once on it, and then use the Ramp Colors button to automatically assign colors in the spectrum above and below the selected class.

✓ **TIP:** *Hold down <Control> and single click on particular colors in a color ramp; then click on the Ramp Colors button to force the new color ramp to pass through those colors. After the color ramp has been set, you can click and drag colors within the Legend Editor into any order you want.*

✗ **WARNING:** *The colors you see on the screen may not match the colors you end up with on the printed page. See Chapter 5 for more information on assigning colors and color matching between the computer screen and the printed page.*

Graduated Symbol

The Graduated Symbol legend type applies to both point and line themes, and is more effective in symbolizing these data types than the Graduated Color legend. Graduated Symbol legends work well with numeric data such as population sizes of cities (represented by points) or classes of highways (represented by lines), from two-lane gravel roads to multi-lane interstates.

When you select Graduated Symbol from the Legend Type drop-down list, shown in the following illustration, you are provided choices similar to those of Graduated Color. You must choose a Classification Field to base the legend on; however, unlike Graduated Color and polygons, it is not necessary but merely optional to normalize the data with point and line data.

ArcView will assign a default point or line symbol in the Legend Editor's tabular display, broken up into the five default classes. Classes can be added or deleted, or sorted with the Legend Editor buttons. In the lower right corner of the Legend Editor, you can choose a Size Range for the symbols, by assigning a minimum and maximum point size, as shown in the following illustration. ArcView assigns the minimum and maximum to the first and last classes and interpolates the point sizes for the intervening classes.

Legend Editor showing the Graduated Symbol legend type.

> ✗ **WARNING:** *If you are using an Equal Interval classification scheme,
> you will get a direct one-to-one correspondence between point sizes and
> class symbols. If you are using any other classification scheme, you may
> get symbol sizes that do not proportionally reflect the number used by
> each class. In this case, you should individually adjust each symbol size
> to more accurately reflect the numeric data.*

Rather than changing the symbol for each class individually, you can
double click on the Symbol box in the lower left corner of the Legend Editor to change all symbols at once. To create your own customized Graduated Symbol ramps, select a symbol for the first class and a
symbol for the last class in the Legend Editor's tabular display and
click on the Ramp Symbols button. ArcView will automatically inter-
polate the point size between the two classes.

> ✓ **TIP:** *An Advanced option, shown in the facing illustration, is available
> in the Graduated Symbol Legend, which allows you to set a Reference
> Scale for both point and line themes. When a reference scale is set, the
> size range of class point sizes will appear exactly at the scale. If the ref-
> erence scale is 1:24,000 and your size ranges are 6 to 18 points, your
> symbols will appear on screen (or on paper if printed at the same scale)
> at 6 to 18 points. If you then zoom in to a larger scale, say 1:2,000,
> the symbols will appear as larger point sizes of 12 to 36 points. If you
> zoom out to 1:100,000, the symbols will appear as smaller point sizes.*

Also included in the Advanced dialog box for point themes is the
ability to rotate the point symbol based on another attribute. The
classic example of this would be wind speeds and directions from
weather stations. The size of the symbol will represent the wind
speed, whereas the rotated direction of the symbol will represent the
wind direction.

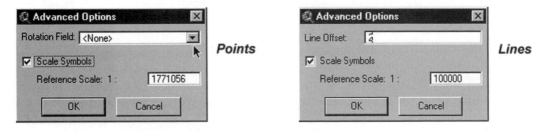

Advanced option in the Graduated Symbol legend type.

The Advanced dialog box for line themes allows you to offset a line from its original location by any number of points. This is very helpful in offsetting roads in one line theme from rivers in another theme. Many roads closely follow rivers, and at some map scales it would appear that the road was in the river. By offsetting the roads by 4 points, the road symbol will correctly parallel and not overprint the river symbol.

Unique Value

The Unique Value legend type assigns a unique symbol (for points and line themes) or a unique fill (for polygon themes) for each class in the Legend Editor's tabular display. This is a useful legend type for displaying categorical data, such as political divisions (e.g., states or counties), or vegetation types, such as evergreen or deciduous forest types.

When you select Unique Value from the Legend Type drop-down list, shown in the following illustration, a Values Field selector becomes available. Once a Values Field is selected from the theme's attribute table, a unique symbol is randomly assigned to the number of records or classes in that field. Using the theme representing the fifty United States, the Values Field can be set to State Name, and each of the fifty states will be filled with a different color. This is useful for showing the differences between state locations, but if a numeric Values Field was chosen (such as 1990 state population), the resulting legend would be a meaningless jumble of different colors, and would not show a progression of colors (such as available with Graduated Color legends) for population sizes.

Legend Editor showing the Unique Value legend type.

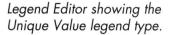

> ✓ **TIP:** *For complex legends with dozens of unique values, copy and paste the theme you are working with in the view, turn the symbol off in the Legend Editor, and label text for polygons. This works well with themes that incorporate a lot of unique colors, such as vegetation, geology, or land-use types.*

Single clicking on the Random Symbols button will globally change both the symbol and the background color for all point or line symbols in the Legend Editor. If the Random Symbols button is used with a polygon theme, the fill for each symbol will randomly cycle through the patterns (not the colors) available for filling polygons. Patterns can be difficult to print and to discern for small areas, and should always be used with much attention paid to their legibility in the map display and the final map.

Single clicking on the Random Colors button will cause ArcView to cycle through four means of randomly determining color for the number of classes displayed in the Legend Editor. ArcView splits the color spectrum up into the number of classes assigned in the Legend Editor, and then varies the saturation and value of each class. ArcView has nine predefined color schemes available from the Color Scheme drop-down list at the bottom of the Unique Value Legend Editor. These are listed in table 3–2, which follows.

Table 3-2: Color Schemes for Unique Value Legends

Color Scheme	Color Ranges
Bountiful Harvest	Browns and tans
Pastels	Light, muted, tasteful colors
Minerals	Dark, highly saturated colors
Fruits and Vegetables	Bright, highly saturated colors
Cool Tones	Blues and greens
Warm Tones	Reds and browns
Autumn Leaves	Dark, highly saturated browns, reds, and yellows
Equatorial Rainforest	Dark, highly saturated greens and blues
The High Seas	Dark, highly saturated blues

These color schemes provide a quick way of symbolizing your data in an appropriate manner consistent with the type of data and the map purpose. Color schemes such as Minerals and Fruits and Vegetables are useful for emphasizing the difference between adjacent areas, whereas Pastels or Cool Tones are more appropriate for polygon themes used as background elements in your map display.

Of course, the different color schemes are limited by the color range of your computer monitor, video card, and the device your map will be printed on. For best results, set your monitor to true color, and calibrate your printer with the color values from your monitor.

> ⊷ **NOTE:** *Color models—including hue, saturation and value, and color calibration—are discussed in detail in Chapter 5.*

Dot Density

Dot maps are a traditional cartographic tool for polygon themes where points are used to represent the value of a particular numeric attribute, and are placed near where that attribute may occur in a polygon. ArcView's Legend Editor provides the cartographer with a Dot legend type for polygon themes, which produces a dot density map. Used properly, dot fills can be more accurate in symbolizing the distribution of a phenomenon across an area than similar color or pattern fills.

Creating a Dot density legend requires at least three steps. After selecting Dot as the legend type, shown in the following illustration, you must choose a numeric attribute for the Density Field. Next, decide how much of the numeric attribute should be represented by each dot. Last, click on the Calculate button before clicking on Apply to see the results in the map display. The Dot legend type gives you the option of choosing a field to normalize the data on, and three boxes that determine the size and type of the dot symbol, the background color and outline of the polygon, and what to use as a null symbol, if anything.

Legend Editor showing Dot legend type.

There are several important considerations when making a dot density map. At the start, it is good practice to determine the extent and scale of the proposed map in the view before calculating a Dot legend type. Once the map scale and extent is set, the ArcView Legend Editor will determine the spacing of the dots based on screen units to give an "optimized density value" for the dot map. Because ArcView uses a random number generator to fill the polygons with dots, the dot fill pattern will change each time the map display in the view is panned or zoomed in or out.

Likewise, the larger the dot size (measured in points), the fewer dots ArcView will place in a polygon. You need to experiment with different dot sizes used at different map scales to arrive at the desired effect. In addition, the size of the polygons in a particular theme can have strikingly different effects when the Dot legend type is applied.

✓ **TIP 1:** *If the Dot legend type does not show any dots in the view when you click on the Calculate button, try zooming in and calculating again with a different value per dot.*

✓ **TIP 2:** *For more realistic dot density maps, use a polygon theme with smaller polygons to represent the dot density distribution in a theme with larger polygons.*

A good example is trying to represent population density by state. If the Dot legend type is applied to a theme of the state of Washington, dots (representing 20,000 people a piece) are randomly distributed within its boundaries. This does not realistically represent population distribution in Washington, which has two large clusters of people: one in the Puget Sound area around Seattle in the west and a smaller one around Spokane in the east. The following illustration shows two types of dot density maps representing population distribution.

You can show this by adding the counties theme for Washington, and symbolizing it as a Dot legend type with the background symbol outline turned off. Then place the state theme on the bottom of the Table of Contents symbolized as a Single Symbol legend type. Now the dot map shows concentrations of population in the west and east of the state. For an even more realistic dot map, substitute a census tracts theme for the county theme, and set each dot equal to 3,000 people.

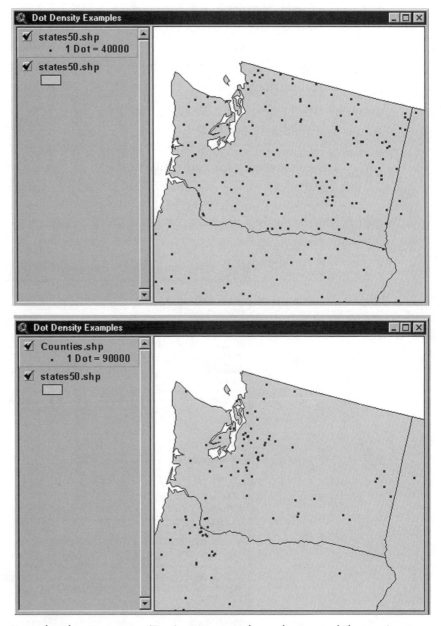

Two dot density maps: (Top) using state boundaries and (bottom) using counties to represent state population distribution.

Chart Symbol

The Chart Symbol legend type, shown in the following illustration, is a type of proportional symbol map that allows you to symbolize several attribute fields at the same time within one theme. The previously described ArcView legend types can symbolize only one attribute at a time within a theme. The first attribute field is symbolized by the size of the bar or pie chart, whereas the second attribute is symbolized by the individual columns of the bar chart or the individual slices of the pie chart. This legend type works well with many types of numeric data, such as demographic information or weather data.

Legend Editor showing the Chart legend type.

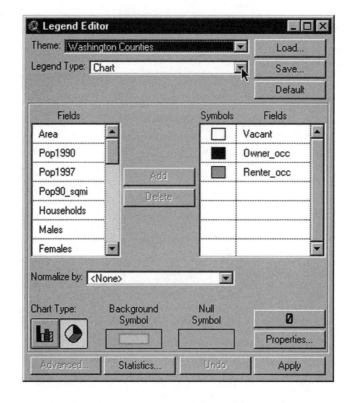

When Chart is chosen from the Legend Type drop-down list, you are presented with a list of fields to add to the Chart Symbol legend. Make sure to add only fields that have related attributes and the same type of data (i.e., all raw counts or all percentages) to the Chart Symbol legend. Once the fields are added to the selection table on the right side of the Legend Editor, the symbols for each field can be selected by double clicking on them to bring up the Color palette. Several other options become active, including the normalization drop-down list, the Background Symbol and Null Value boxes, and the Bar or Pie chart button.

More importantly, you can click on the Properties button to choose a Size field with a maximum and minimum size that will scale the size of the individual bar or pie charts. Examine the attribute field you are basing the Chart Symbol size on and try several ranges in order to arrive at a range that most accurately depicts the range of data for that attribute. For example, if population is chosen for the Size field, but the range of difference for populations among the features in the view is minor, you do not want the symbol size differences to be very large, which would give the impression that the population differences were great.

There are limitations with the Chart Symbol legend. The location ArcView places the Chart symbol in each feature in the view cannot be changed; nor can the feature order in the view. This may cause some symbols to obscure other Chart symbols, and the only way to make them more readable is to reduce the maximum range of the Size field.

Summary

- -

The ArcView Legend Editor provides you with the ability to create six different legend types, but it is important to understand which legend type is appropriate for what type of data. Although powerful, the Legend Editor will not always automatically display your data in a cartographically correct manner. It is up to the cartographer to know the data well and to use that knowledge to display it on the map in a way that informs rather than misleads the map's audience.

Chapter 4

Classification in the Legend Editor

0   Miles   500

One of the more crucial steps in the preparation of a map in ArcView is the step of choosing a classification method. Classification is the process of placing data with similar values into groups, in order to make the data easier to understand. If you choose a Graduated Symbol or Graduated Color legend type for numeric data in the Legend Editor, you are faced with a choice of five classification methods: Natural Breaks, Quantile, Equal Interval, Equal Area, and Standard Deviation. What method you choose depends on the numerical distribution of data in a field in your attribute table, and what aspect of that distribution you want to show cartographically.

The purpose of using a classification scheme for your legend is to make a point about your data's geographic distribution in a clear manner to the readers of your map. It takes some knowledge of the data (from the metadata file) and some time exploring classification schemes in the Legend Editor to achieve a useful and easily understandable legend classification in the final map. Remember that the classification method and number of classes you choose appear first in the view's Table of Contents and eventually in the legend in the final map produced from the layout document. If your final map is to have value, it is important to make the legend understandable.

Getting Started with Classification: Numerical Data Distribution

Before investigating particular classification methods, briefly examine some ways numerical data can be distributed. It is often helpful to chart data in order to visualize it better. Unfortunately, ArcView's charting features provide little help in understanding the specifics of data distribution. Therefore, it is better to export selected fields of an attribute table to software that has more powerful graphing capabilities, such as Microsoft Excel. Within ArcView, you can visually examine distribution of features by sorting by ascending or descending values in a field in an attribute table.

Numerical data can be evenly distributed, unevenly distributed, continuously distributed, or normally distributed. The following illustrations show four types of data distribution, with idealized classification methods for each.

Bar chart of unevenly and evenly distributed data.

Bar chart of continuously and normally distributed data.

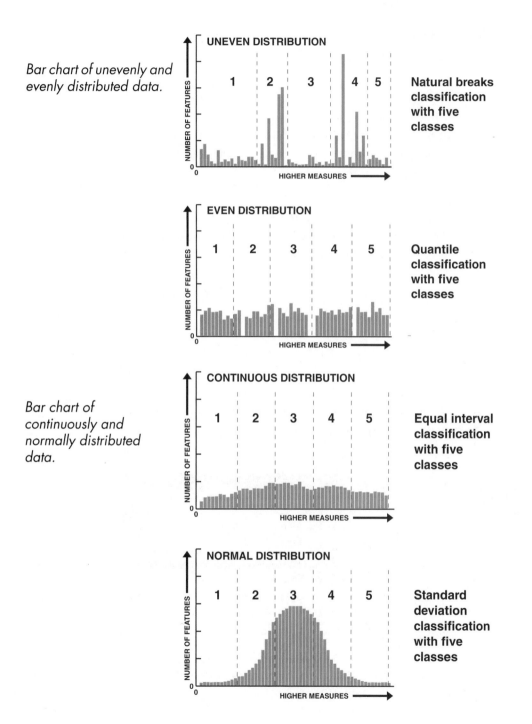

UNEVEN DISTRIBUTION

Natural breaks classification with five classes

EVEN DISTRIBUTION

Quantile classification with five classes

CONTINUOUS DISTRIBUTION

Equal interval classification with five classes

NORMAL DISTRIBUTION

Standard deviation classification with five classes

Classification Methods

ArcView's five classification methods are Natural Breaks, Quantile, Equal Interval, Equal Area, and Standard Deviation. Each has its strengths and weaknesses. The trick is to choose one that shows an important feature of your data in a way that is easy for the audience of your map to understand.

You begin by choosing Graduated Symbol or Graduated Color legend types in the Legend Editor, and specifying a Values field. ArcView defaults to a Natural Breaks classification method with five classes.

You can change the classification by single clicking on the Classify button to bring up the Classification dialog box, shown in the following illustration. Here the method is easily changed in the Type dropdown menu, and the number of classes can be altered from the default of five to anywhere from one to 64. The number 5 is actually a good default value for classes because research has shown that people have a difficult time understanding greater than seven or eight classes on a map.

Legend Editor and the Classification dialog box.

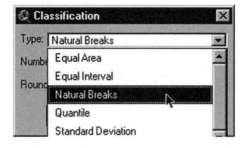

You can add, subtract, and hide classes (using "no data" symbols) and assign legend labels by replacing numeric labels with text labels. The following table lists the classification methods and ranks their performance based on various criteria.

Table 4-1: Advantages of Various Classification Methods[1]

Classification Type	Natural Breaks	Quantile	Equal Interval	Equal Area	Standard Deviation
Considers distribution of data along a number line	VG	P	P	P	G
Ease of understanding concept	VG	VG	VG	G	VG
Ease of understanding legend	P	P	VG	P	G
Legend values match range of data in a class	VG	VG	P	P	P
Best data distribution for classification	Data not evenly distributed	Data evenly distributed	Data continuously distributed	Data continuously distributed	Data normally distributed

VG = very good   G = good   P = poor

[1] Adapted from T. A. Slocum, *Thematic Cartography and Visualization*, Prentice Hall, 1999.

Natural Breaks Classification

The Legend Editor defaults to a Natural Breaks classification in five classes. ArcView uses a clever algorithm, called the Jenks Optimization algorithm, to group values within a class, resulting in classes of similar values separated by breakpoints. This method works well with data that is not evenly distributed and not heavily skewed toward one end of a distribution.

Although Natural Breaks, shown in the first of the following illustrations, is a good choice for a default method of discovering patterns in your data, it can frequently result in legends that are difficult for the map reader to understand. Strange number ranges can appear in the legend. You can try to change this by interactively tweaking the number ranges in the Legend Editor table. However, if you tweak them too much, the legend can become an uneven mess. The second of the following illustrations shows a map with a resultant legend.

Legend Editor and View with a Natural Breaks classification.

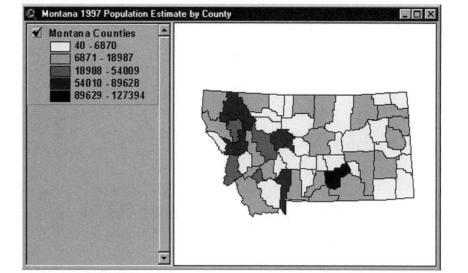

Legend showing number ranges.

A better way of handling peculiar number ranges is to change the Labels in the Legend Editor from the numeric labels to a text label. This way, number ranges difficult to understand can be labeled as low, medium, and high values, which are much more easily and quickly comprehended by the map reader.

Quantile Classification

With a Quantile classification, shown in the first of the following illustrations, equal numbers of features (and corresponding records in the attribute table) are placed in each class. For example, if you have a theme with 20 features and you apply a Quantile classification, Arc-View will, by default, place four features into each of five classes. This classification method works best with evenly distributed data. The second of the following illustrations shows a resultant Quantile classification map.

Default quantile classifications can result in very misleading legends. In order to have an equal number of features in each class, features with greatly different values can be placed in a single class. One solution is to manually increase the number of classes, in an attempt to separate dissimilar values.

Quantile classification in the Legend Editor.

Quantile classification map.

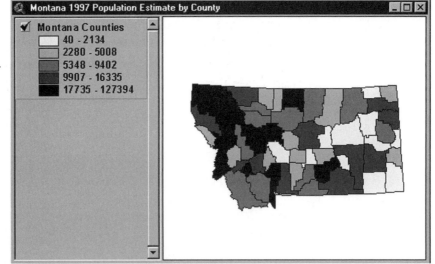

Equal Interval Classification

Using an Equal Interval classification, shown in the first of the following illustrations, ArcView divides the total range of feature values, from maximum to minimum, into five equal subranges. This creates an easy to understand legend and works best with continuously distributed data, such as precipitation amounts across North America. The second of the following illustrations shows a resultant Equal Interval classification map.

Legend Editor with an Equal Interval classification.

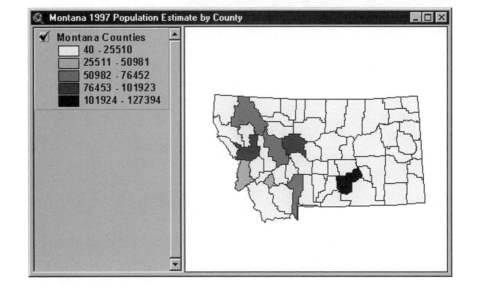

Equal Interval classification map.

The downside of this method is that if it is applied to certain data types, the legend can end up with classes with few, if any, map features contained within them. If you attempt to tweak the class ranges by editing the values in the Legend Editor, you fundamentally change the concept of Equal Interval classifications.

You are also faced with the problem of presenting a legend with classes with no features or data in them. Should you delete that class or show it? One solution is to keep the class in the legend, but note that it contains no features on the map.

Equal Area Classification

The Equal Area classification, shown in the first of the following illustrations, is a very specialized ArcView method that divides the range of features in a polygon theme so that each subrange contains basically the same area. This method is useful in a limited number of situations, such as apportioning sampling areas so that every observer gets an equal area. The second of the following illustrations shows a resultant Equal Area classification map.

*Legend Editor
with an Equal
Area
classification.*

*Equal Area
classification
map.*

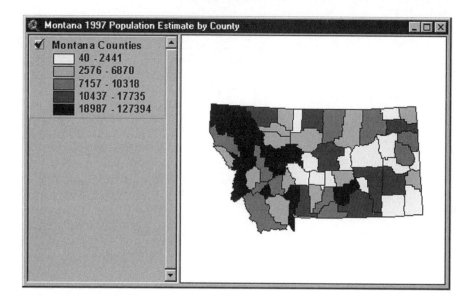

Standard Deviation Classification

In the Standard Deviation Classification method, ArcView calculates the mean value of a field in a theme, and then divides the range of feature values by several deviations above and below the mean. This method works well with normally distributed data; that is, data that fits a standard bell curve, with most feature values falling in the center, and few high and low feature values appearing above and below. The first of the following illustrations shows a Standard Deviation classification. The second of the following illustrations shows a resultant Standard Deviation classification map.

Legend Editor with a Standard Deviation classification.

*Standard
Deviation
classification
map.*

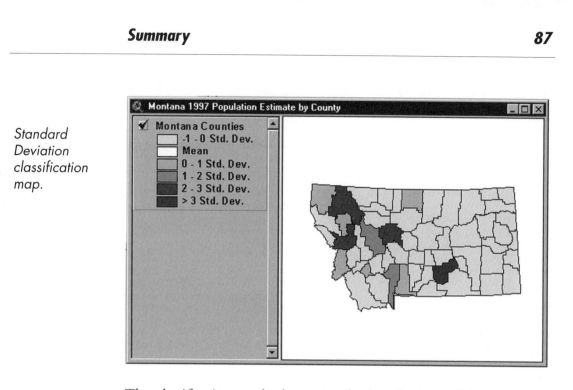

The classification methods previously described are all being applied to the same field in one theme: 1997 population estimates for Montana's fifty-six counties. Each classification method results in a completely different map, with some more difficult to understand than others. When using classifications, make sure to explore your data by applying several methods and choosing one that shows a significant feature on your map.

Summary

The ArcView Legend Editor provides five means of classifying numeric data. What method you choose depends on the numerical distribution of data itself and how you want to display that distribution visually. The Legend Editor provides a powerful tool for exploring the relationships within your geographic data, but a good grounding in statistics is helpful when choosing a means of classification for that data.

Chapter 5

Palettes in the ArcView Legend Editor

Manipulating Markers, Colors, and Patterns in the View

Buried several mouse clicks into the ArcView Legend Editor is an important set of six windows that control most of the appearance of features in the view and therefore your layout and final map. These six windows (or "palettes") are collectively called the Symbol Window and include the Fill Palette, Pen Palette, Marker Palette, Font Palette, Color Palette, and Palette Manager. These palettes are shown in the following illustration. This chapter examines the use of all of these, except for the Font Palette, which is important enough to be the focus of its own chapter, Chapter 6, Typography on the Map.

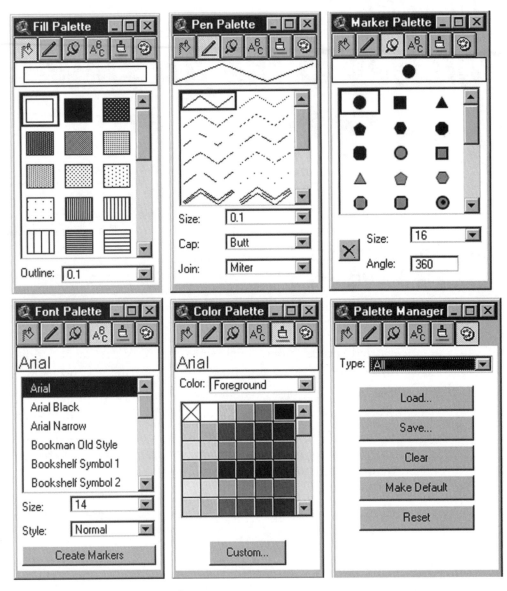

The Symbol Window's six palettes.

Learning the technical details of how to use these palettes is relatively easy, being mostly a matter of pointing and clicking. Knowing how to apply the visual tools of shape, size, texture, hue, and value in the Fill,

Pen, Marker, and Color palettes is more difficult, but becomes second nature as you gain experience in making maps with ArcView. Throughout the discussion of how each palette works, you will see how to apply cartographic design principles with the tools offered in each palette.

Introducing the Palettes

The Symbol Window can be accessed by one of three methods: (1) by double clicking on a symbol within the Legend Editor, (2) by using Show Symbol Window from the Window menu in the view, or (3) by using Control P on the keyboard. The Fill palette appears at first on the screen, and the other five palettes are accessed by pressing their respective buttons, which are displayed in a row below the Fill palette's title bar. Palettes are defined in the ArcView help file as "a collection of symbols or colors stored in a file that you can load or create." Table 5-1, which follows, summarizes the types of features available in each palette, using the default palette.

Table 5-1: Symbol Window Palettes

Palette Name	Basic Features	Additional Features
Fill Palette	47 default polygon fills	Polygon outline weights in points
Pen Palette	28 default line types	Line widths in points
Marker Palette	47 default marker types	Marker sizes in points
Color Palette	60 default color swatches	Custom colors using the Hue, Saturation and Value color model
Font Palette	Several dozen font types	9 specialized ESRI fonts that can be made into markers
Palette Manager	5 palette management options	Load and delete additional palettes

Cartographic Design Considerations at Start-up

The Symbol Window and its six palettes can present an overwhelming number of design choices for the beginning cartographer, resulting in a mishmash of colors and symbols on the final map and a confused, if not dazed, map user. Several cartographic design principles apply to the use of the palettes and are good to keep in mind as you craft the map.

- For all but the most complex reference maps, keep the design simple.

Just because you can print hundreds of colors and dozens of symbols on a map does not mean you have to, or that the map will be understandable when you do! Maps are always a simplification of real-world geography, and a good cartographer simplifies the complex geography to make a point clearly to his or her audience.

- Consider the type of map you are crafting and the type of map product that will result.

This principle has been mentioned often before, but in no case is it more important than in the use of colors and patterns, and markers and line symbols. Different types of maps require different amounts of color and symbol complexity. Thematic maps showing the distribution of one or two map features need the simplest color schemes, whereas general reference maps or maps using different types of raster imagery will require the most complex.

A full color map published in a book or atlas can utilize dozens of colors and symbols, whereas maps meant to be viewed on the Internet need to have simple color schemes. If your map will be reproduced on a photocopier or reprinted as black and white in a book or newsletter, your color scheme needs to be simpler still. The following illustration shows the progression from simple to complex types of maps, with their associated symbols and colors.

The Use of Color, Patterns, and Markers

Color and symbol complexity in different types of maps.

Type of Map	Amount of Color and Patterns	Number of Markers
Simple Thematic Maps —Sales Maps	Fewer than a dozen colors one major color progression	Less than half a dozen types of markers; one marker hierarchy
Grayscale Aerial Photography		
Complex Thematic Maps —Mutlivariate —Thematic Maps	Dozens of colors; two major color progressions	A dozen types of markers; a couple of marker hierarchies
Color Aerial Photography		
Mutlispectral Satellite Imagery		
Complex Reference Maps —Atlas Maps —Shaded Relief Maps —Vegetation Maps —Land-cover Maps	Hundreds to thousands of colors; multiple color progressions	Dozens of types of markers; multiple marker hierarchies

- Always strive to establish a visual hierarchy for symbols, lines, and colors on the map.

Small circles for cities with small populations and larger circles for cities with large populations; thin lines for two-lane country roads and thick lines for major interstate highways; a light colored polygon for low sales figures and a darker colored polygon for high sales figures: These types of visual hierarchies organize data for the map reader and make the map quickly understandable.

- Always check the symbol legibility on the intended map product before using it on the entire map.

Many pen and point symbols look perfectly intelligible when you are zoomed into a view, but will not print clearly on some printers, or become invisible when used as a GIF on a web page. Save design time by trying a draft with several symbols you want to use on the map before investing a lot of time in production.

⟿ **NOTE:** *Remember the golden rule: If the map user cannot read it quickly, you should not use it on the map.*

Using the Palette Manager

The Palette Manager permits the loading and unloading of specialized palettes for the Fill, Pen, Marker, and Color palettes, dozens of which come with the ArcView program (with hundreds more available on the Internet). With certain other software programs or additional Arc-View extensions, you can create your own specialized palettes and use them in lieu of the default palettes. Table 5-2, which follows, lists ArcView specialized palettes, with notes on their features.

Table 5-2: ArcView Specialized Palettes

File Name	Features
Fill Palette	
carto.avp	48 vector fills with transparent backgrounds, based on ArcInfo shadeset carto.shd
hatch.avp	60 vector fills; 6 patterns with 10 progressions
raster.avp	33 raster fills used for Unique Value fills by ArcView
Pen Palette	
carto.avp	28 pen symbols
forestry.avp	10 line symbols used on U.S. Forest Service maps

File Name	Features
geology.avp	168 line symbols for major geological features
transp.avp	34 highway and railroad lines
weather.avp	16 line symbols representing weather fronts
Marker Palette	
amfm.avp	24 markers for facility management
colormrk.avp	50 markers with changeable foreground colors
arrows.avp	25 directional arrows
crime.avp	63 markers for crime mapping
envtl.avp	58 markers based on international oil spill cleanups
hazmat.avp	19 symbols for hazardous material warning labels
icons.avp	54 symbols from ArcInfo *plotter.mrk* and *template.mrk*
mineral.avp	41 mining symbols
municipl.avp	88 municipal activity markers
north.avp	11 north arrows
oilgas.avp	90 USGS oil and gas activity markers
realty.avp	43 symbols used in real estate maps
usgs.avp	43 markers from USGS topographic maps
water.avp	50 USGS water resource management symbols
forestry.avp	21 markers from U.S. Forest Service fire maps
geology.avp	108 USGS geological symbols
raster.avp	135 markers used as default by ArcView
transp.avp	89 highway and street signs
weather.avp	185 USGS weather symbols
Color Palette	
c256.avp	256 colors based on MS Windows 256 color display
colornam.avp	112 colors from ArcInfo shadeset *colorname.shd*
hardware.avp	17 high contrast colors from ArcInfo shadeset *color.shd*
rainbow.avp	259 color ramp based colors
rgbtext.avp	626 colors based on UNIX X windows good on PCs
safety.avp	216 non-ditherable colors safe for web-based maps

⇢ **NOTE:** *Most of these palettes, except for the Color palettes, are shown in the appendix of the manual that comes with the ArcView software package, called "Using ArcView GIS."*

Palette Management Options

The Load button allows you to load new palette files (files with an *.avp* suffix) to one or all of the default palettes. Make sure to scroll down in a palette window to see the new palette, which is typically added at the bottom of the existing default palette. The following illustration shows the loading of new palettes with the Palette Manager.

Loading new palettes with the Palette Manager.

The Save button will save a new markerset produced by the Create Markers option in the Marker palette or an imported ARC/INFO shadeset or lineset as a new ArcView palette *(.avp)* file. The Clear button will remove any added palettes.

If there are no additional palettes loaded, the Clear button will remove the default palettes. You will know if this is the case when all palettes show up as blanks! To restore the default palettes, use the Load button to locate the *default.avp* palette file in the ArcView installation directory, which on Windows PCs is typically *c:\esri\av_gis30\ arcview\symbols\default.avp*.

Unfortunately, ArcView does not remember any new palettes loaded when a project is closed and reopened, but reverts to *default.avp*. If you find yourself using several new palettes in much of your work, it is worthwhile to create a new default palette.

Before clicking on the Make Default button in the Palette Manager, go to the *c:\esri\av_gis30\arcview\symbols* directory and rename *default. avp* as *defaultorig.avp*. Now when you use the Make Default button to create a new *default.avp* file, it will not overwrite the original system setting. Using the Reset button will automatically unload any new palettes, and revert to the *default.avp* palette file.

Color and Fill Palettes and the Use of Color

You should explore the use of the Fill and Color palettes together, because they work in tandem to control the appearance of any polygon themes in the view. The Fill palette controls the appearance of polygon features in the view by managing area fills, using solid colors, transparent fill, or pattern fills (see the following illustration). In addition, the line weight of the polygon's outline can be set in this palette. When the Fill palette first appears, a solid fill is chosen as the default. This will result in polygons filled with solid colors as selected from the Color palette.

The Fill palette options that control the Color palette.

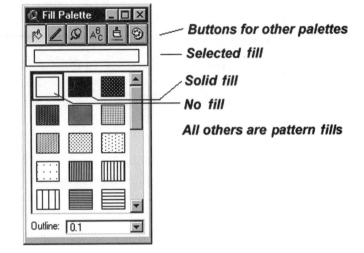

✓ **TIP:** *In the Fill palette, ArcView defaults to a .1-pt line weight for polygon outlines, which is a very thin line weight. Change this weight to at least .5 pt or greater if you want your polygon outlines to be visible on a printed map.*

Colors and Fills

The Color palette provides 60 color patches as a default option for applying to points, lines, polygons, and text, and permits the creation of custom colors, using the hue, saturation, and value (HSV) color model. In general, lighter colors appear on the left side of the palette, whereas darker colors appear on the right, giving the user a quick choice of a simple color progression from left to right. The following illustration shows the Fill and Color palettes and the Specify Color dialog box.

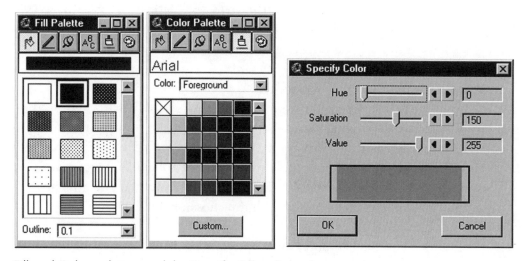

Fill and Color palettes, and the Specify Color dialog box.

✓ **TIP:** *Color progressions are very useful in thematic mapping to show differences in magnitude, in that map readers typically view lighter shades of one color to signify "less" of something, whereas darker shades of the same color indicate "more" of something. Use completely different colors to show differences in kind, not magnitude.*

In regard to the previous tip, for example, in the Legend Editor, choose a Values field that is a numeric field, and use light red colors for low numbers and dark red colors for high numbers. Also for example, when using a Graduated Color legend type, use one of the pre-made color ramps.

✓ **TIP:** *Another good use of color in thematic mapping is for enhancing figure-to-ground contrast. You can show the distribution of a feature (the figure), depicted with darker colors, across a map containing light colors in the background (the ground). When done right, the darker features will "float" above the light background, achieving good figure-to-ground contrast.*

In general, thematic maps will use darker, more intense colors to achieve their communication goals than will reference maps, which utilize lighter shades in order make complex detail easily visible to the reader.

At the top of the Color palette, below the Palette buttons, is a window showing the currently selected color swatch. Below that is the color drop-down menu, which offers a choice of Foreground, Background, Outline, or Text colors. For solid fills, Foreground Color is the most commonly used option, resulting in a polygon, line, or symbol completely filled with the chosen color.

The Background Color option is used with some, but not all, pattern fills and some symbols, which consist of foreground and background objects. The Outline Color option allows you to make the outline of a polygon different from the foreground fill. Use the Text Color option for coloring the map's text labels something other than black. Below the default color swatches is the Custom button, which brings up the Specify Color dialog box, which allows you to specify a color as visualized in the HSV color model.

Color Models in ArcView

In the printing and publishing world, several types of color models can be used on computers to represent color. ArcView uses an HSV (hue, saturation, and value) color model as its standard to generate colors, but a special ArcView extension gives you access to the more common RGB (red, green, blue) and CMYK (cyan, magenta, yellow, and black) models.

The HSV color model is called a "user oriented" model because it relates to the way people perceive color. Hue is what we commonly refer to as color, representing the various wavelengths of light that make up each color, primarily values of red, green, or blue. Saturation refers to the amount of gray added to a hue. A very saturated hue has very little gray, whereas a desaturated hue has a lot of gray. Value relates to the amount of lightness or darkness within a color. Amounts of hue, saturation, and value are measured from 0 to 255. An HSV reading of 0,0,0 is black, whereas a reading of 0,0,255 is white.

The RGB and CMYK models are called "machine" or "hardware" oriented, because they are based on the way computer monitors and most printers use color. The RGB color model refers to the three electron guns in a cathode ray tube display that make up computer monitors and television screens. Each of the three guns hits the inside of the computer screen and excites different phosphors, which result in different additive colors on the screen. Amounts of red, green, and blue are measured from 0 to 255. An RGB value of 0,0,0 is black, whereas an RGB value of 255,255,255 is white.

The RGB color model is typically used by computer photo editing software, and is appropriate to use when designing maps intended for use on the Internet, where most graphics will be viewed on a computer monitor. The CMYK color model is frequently called the process color model, referring to the four-process ink colors of cyan, magenta, yellow, and black. The values of each ink range from 0 to 100, with a CMYK value of 0,0,0,0 equaling white and 0,0,0,100 representing black.

CMYK process colors are the most common method of specifying ink colors in the publishing industry, whereas HSV colors are never used in the printing process. A host of books are available with CMYK color charts, allowing the user to see what a particular CMYK color will look like when printed on paper, and giving the user the CMYK values to dial into their software. This cannot be done in ArcView without using a special extension, called the Color Picker Extension, discussed in the following material.

Using the HSV, RGB, and CMYK Color Picker Extension

Because ArcView uses the HSV color model, which is not used in printing, it is necessary to use another software extension to specify RGB or CMYK colors. Luckily, ESRI has done just that by producing the HSV, RGB, or CMYK Color Picker extension.

↔ **NOTE:** *This extension can be found on the World Wide Web on the Arc-Scripts page at the address* http://www.esri.com.

To use the extension, download the compressed file from the web and uncompress it (easily done using WinZip on the PC, from *http://www.winzip.com*). Make sure ArcView is not running, and copy the actual extension file, named createcolors.avx, into the ArcView installation directory for extensions, usually found at *c:\esri\avgis_30\arcview\ext32.*

The extension appears as a multi-colored icon to the right of the Help icon on the Button bar when the ArcView Project Window is active. Clicking on the multi-colored icon brings up the HSV, RGB, or CMYK Color Picker dialog box, shown in the following illustration.

The Color Picker operates from this single window. The box at the top shows the current color selected, below which are displayed the three common color models: HSV, RGB, and CMYK. To the right of the color models is the default ArcView Color palette, with blank spaces on the bottom for new color swatches. Table 5-3, which follows, compares color model standard values and ArcView color values.

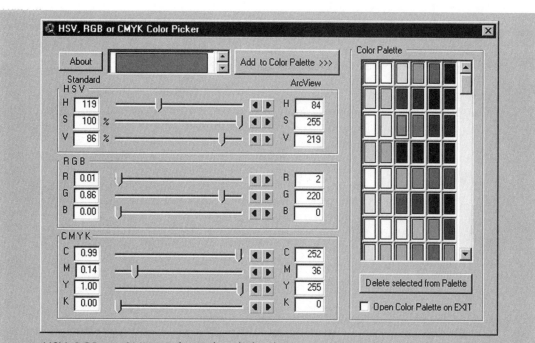

HSV, RGB, or CMYK Color Picker dialog box.

Table 5-3: Color Model Standard Values Versus ArcView Color Values

Color Model	Standard Color Value	ArcView Color Value
HSV Model		
Hue	0–360	0–255
Saturation	0–100%	0–255
Value	0–100%	0–255
RGB Model		
Red	0–1.00	0–255
Green	0–1.00	0–255
Blue	0–1.00	0–255
CMYK Model		
Cyan	0–1.00	0–255
Magenta	0–1.00	0–255
Yellow	0–1.00	0–255
Black	0–1.00	0–255

The user can type in color values in the standard boxes on the far left, or use the sliders to select a color. As one slider or value is dialed in, all sliders and values change to show how that color is represented in each of the three color models. Likewise, you can double click on a color in the palette to the right, and the color's values will appear in the color models to the left. If you check the box on the lower right, the Symbol Window's Color palette will open when you exit, showing your new color swatches added to your default palette.

This is a very powerful method of creating your own color palettes and swatches. You can experiment and enter your own color swatches, or more commonly, create specialized color palettes from color values found in reference books, technical articles, and presentations.

When you are finished using the Color Picker, make sure to save your new Color palette by using the Save command in the Palette Manager. The new palette must have an *.avp* suffix, but once it has been saved to your hard drive, it can be loaded into any other ArcView project.

Calibrating Your Computer Monitor to Your Printer

WYSIWYG, or "What you see is what you get," continues to be the holy grail of computer graphics, and in no instance is it more crucial than matching the color on your computer monitor to the color on your color printer. Computer graphics professionals in publishing operations or advertising firms will use specialized high-end software and pricey printers to ensure a close match between screen and paper, but most ArcView users do not have access to these luxuries.

Luckily, ESRI has provided an ArcView project called *calibrat.apr* that contains a layout document offering CMYK color combinations and progressions that can be printed on almost any color printer. Using *cali-*

brat.apr and a few tricks can make the color matching job a lot faster and easier.

At the start, let your computer monitor warm up for 20 to 30 minutes, which ensures that your display will show the same colors consistently. Then make sure your computer's monitor is set for the highest number of colors your video card will allow. On a Windows 95 (or higher) computer, select Start menu ➡ Settings ➡ Control Panel ➡ Display. In the Display panel, click on the Settings tab and set the Color palette for the largest amount possible. Heftier video cards will allow 65,536 colors or "true" color (32,000,000 colors).

If possible, try to view your monitor and printed map under natural lighting conditions or incandescent light rather than under fluorescent lighting. Most fluorescent lighting will add a distinct blue tint to printed materials, although some fluorescent tubes offer "full spectrum" light that obviates this problem.

Next, find the ArcView project *calibrat.apr*, and open it in ArcView. *Calibrat.apr* is usually found under Samples in the ArcView installation directory. With Windows operating systems, it is *c:\ESRI\AV_GIS30\ ARC-VIEW\Samples\other\ calibrat.apr.*

➥ **NOTE:** *If you cannot find it on your machine, a sample copy is included on the companion CD-ROMs.*

When the project opens, it contains just one document, a layout named Color Test, shown in the following illustration. Open Color Test, which is an 8.5-inch by 11-inch page, and print it to your color printer.

The Color Test layout provided by calibrat.apr

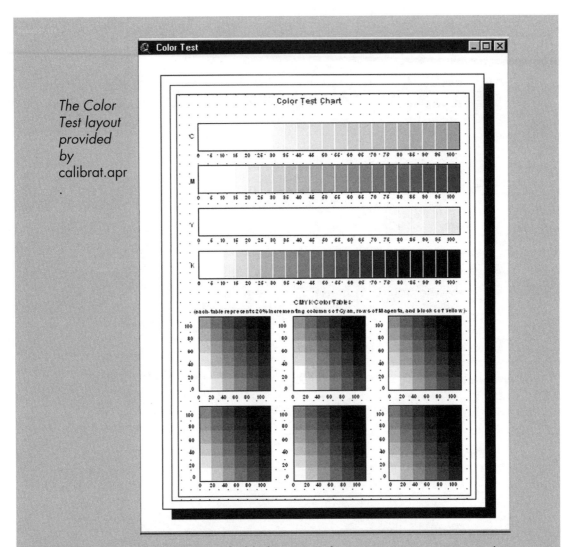

After printing the layout, hold the printed page next to your monitor with the layout displayed and compare screen and printer colors. This is a crude, but fast and inexpensive, means of calibrating the computer screen to your printer's color.

Fill Palettes and the Use of Patterns

- -

The Fill palette presents three basic polygon fill choices for the cartographer: solid fills (the solid black swatch in the middle at the top of the default Fill palette), no fills (the white, empty swatch in the upper left), and pattern fills (everything else in the fill palette). Most commonly, you will click on the solid fill option, and go on to the Color palette to choose an appropriate color for the area on the map. Using pattern fills well in cartographic design is more difficult.

In many ways, pattern fills are a holdover from the days of traditional cartography, with hand-drawn maps using pens and paste-on fills. In those days (as recently as the 1980s!), putting color on maps was a difficult process, and getting good color reproduction on the final copies an expensive proposition. Rather than using color to show differences between areas on a map, you would use press-on black-and-white patterns cut to size for each polygon with an Exacto knife, and jiggered into place. Once you were able to place the patterns correctly, the black-and-white map tended to reproduce well when photocopied.

Today, with inexpensive color printers and color photocopiers, patterns are used far less frequently on maps, being replaced by color coding. This is a good move, because color is a better tool for showing differences in map features than patterns. For example, color is a much better tool when used in a color progression (different shades of one color) to show differences in magnitude for a map feature's attribute, such as population size or sales figures. In addition, patterns need to be used with relatively large areas, because the map reader will have difficulty discerning a pattern filling a small polygon.

However, pattern fills still have a place in digital cartography, especially in the creation of geologic maps, landcover maps, and vegetation maps. In these cases, patterns are used to indicate differences in kind, among different rock types or vegetation communities, not differences in magnitude.

Once you choose a pattern fill in the Fill palette, you can use the Color palette, shown in the following illustration, to change the pattern's foreground and background color. With some pattern fills, such as ArcView's default pattern fills, you cannot make the background transparent in order to see through the pattern to an underlying theme. This is because the default Fill palette is a raster- or bitmap-generated pattern.

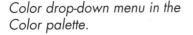

Color drop-down menu in the Color palette.

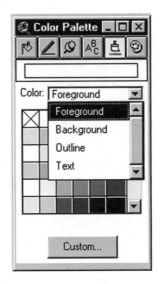

✓ **TIP:** *To create polygon fills with crisp patterns and a transparent background, load the* carto.avp *palette file using the Palette Manager from the ArcView symbols directory.*

Another useful palette is *hatch.avp,* which contains six patterns with several progressions created by increasingly dense patterns. These pattern fills can be used to show differences in magnitude for map features, and are useful if your final map must be produced in black and white only.

Always make sure to print out pattern fills before you commit to using them on a map, to see how they reproduce on your printer. The technique described in the "Printing ArcView Palettes with the Symbol Dump Script" section in this chapter is invaluable in this case.

The Marker Palette and Point Symbology

The Marker palette manipulates marker or point symbols in the view, providing the largest variety of symbols for many types of specialized mapping needs. The default Marker palette features 47 point markers, but hundreds more are available from the ArcView samples directory and from the Internet. Below the default pen window are options for changing marker size (measured in points) and marker angles (measured in degrees). Marker sizes range from 2 pt to 72 pt, but any size or fraction of a size can be manually typed into the size box. Make sure to press the <Enter> key to ensure that the manually entered size is accepted by ArcView.

Many, but not all, markers can be rotated using the Angle option. This feature is useful when you need a particular point symbol to stand out in distinctive fashion, using a flag symbol example to point to a particular location.

Marker symbols have design concerns that involve visual organization and legibility. Just because you have access to hundreds of different types of point symbols does not mean you should use all of them on the same map! Only the most complex maps, such as geological maps, industrial facility mapping, and the like use dozens of symbol sets. Most thematic maps can get by with a handful of different marker types to communicate their message.

> ✓ **TIP:** *Make sure to establish an understandable visual hierarchy. Use different types of markers to indicate qualitative differences (differences in kind). Use differences in symbol size to indicate differences in magnitude. A common example is varying sizes of circles to show the differences in population sizes for cities.*

The legibility of the line work that makes up individual symbols is a chief concern. Many symbols are attractively drawn on the screen, but are not robust enough to hold ink when printed on paper. As with lines and colors, you must try several drafts with different marker types and sizes to see if they reproduce well on the final map.

You can also change the foreground color of the markers in the Color palette. Most marker sets will not permit you to change the outline color (or the background color, if the symbol has one). Keep in mind that most symbols are small compared to the entire map, and subtle changes in colors may not be noticeable on paper, because the markers hold little ink.

A unique tool found in the Font palette, called Create Markers, allows you to change any font in the Font palette to a markerset for use in the Marker palette. Many symbol fonts are available commercially or for free on the Internet and can be made into point symbols with the Create Markers option.

Finding and Creating Specialized Palettes

Extra palettes are available in the ArcView samples directory, and hundreds more are available on the World Wide Web. A good place to start is the ArcScripts page accessed from the ESRI homepage at *http:// www.esri.com*. Once there, search for "palettes" using the page's search feature, or just browse the hundreds of scripts, palettes, and utilities listed for ArcView 3.x.

One of the best ArcView resources on the web is the Unofficial ARC/ INFO and ArcView Symbol Page (see the following illustration), created by Brian Sheahan and found at *http://www.ifas.ufl.edu/~bts/gis/symbols.html*. This page has a wealth of information and links to non-copyrighted specialized palettes and utilities for use with ArcView. There are many unique symbol sets for representing highway shields, public land ownership such as national parks, international symbols for recreation areas, and military symbols.

Of particular interest are the palettes such as Artist, produced by Jim Mossman. The Artist palette loads 800 RGB colors in dozens of progressions to the color palette, giving the ArcView user virtually any color combination you would ever need.

For those cartographers migrating from ArcInfo to ArcView, it is possible to load certain ArcInfo shadeset and lineset files into the ArcView Symbol Window. Use the Palette Manager's load option to load shadesets or linesets, or use *AIshd to Avcolor*, a script from the Unofficial AI and AV Symbol page, to convert ArcInfo shadesets to ArcView palettes.

This page is the result of a query I posted to the ESRI-L and ArcView-L users list in March 1998. From the onset of my career I ran into situations where the existing cartographic symbology was not adequate for particular projects. The result was having to spend hours, days or in some cases weeks creating the the needed symbols. After becoming active in the Users Groups I realized many of my colleagues faced the same obstacles and with the very generous help of others I was able to create this resource.

Symbols & Fonts	Symbol Set Resources & Tools
Frequently Asked Questions	Font Editors & Resources
Guest Book/Suggestions	Icons

Symbols & Fonts

Webpal.zip (2KB)	**RGB Palette** A color palette based on RGB values obtained from a Web Programmer's book to produce "non-dithered" colors. Lot's of good colors not available in other data sets.
Recreate.zip (24KB) Macintosh Users use: **Macrec.zip** (24KB)	**Recreational Symbols** A WINDOWS format TrueType symbol font of a wide variety of recreational symbols. These were created by our graphics department for placement onto park, trail, WMA, and other maps created for the department. This TrueType font can be converted to markers and used for point classification. The recreate.ttf file is great for symbolizing point data sets. To do this you need to convert the font to a marker set using the CREATE MARKERS button in the Text Palette window of the symbol Palette. You can open the symbol Palette using the WINDOW: Show Symbol Window option. Then you can use them to symbolize the point using the legend editor and classification and symbolization windows.
US NPS Symbol Set	**National Parks Service Symbol Set.** Thanks to the generosity of the U.S. National Parks Service, WV, Avenza Software Marketing Inc. was provided with a public domain Adobe Illustrator file (PrksStandard) containing the symbols and pictographs that are used in the creation of National Parks Service maps. The file contains both a Mac font and a TTF font. Copyright © 1997 Avenza Software Marketing Inc.
Newfill.avp (9KB)	This is the default symbol fill palette with one extra fill of the **USGS swamp symbol** at the the end of the list. I first have people clear the fill palette and then load the new one.
GSC Symbol Set (135 Kb)	The **Geologic Survey of Canada Symbol Set** includes shade, line, and marker sets and the necessary fonts. Also included are AML's for plotting the symbol sets. They go some way toward explaining the various symbols, but one might want to refer to the legend on a published GSC map for clarification. Thanks to Doug Hrynyk for convicing GSC to distribute this set.
Road Symbols (16KB)	A highway symbol creator written in Avenue as an extension for ArcView. The extension creates four different black and white road symbols and populates them with values read from a database. The zip file has minimal documentation, the extension and a true-type font. **The Symbol Sets & Extension above were created by:** Tim Loesch, GIS Applications Programmer, MIS Bureau, Minnesota Department of Natural Resources.

Unofficial ArcView Symbol Page.

The Pen Palette and Line Symbology

The Pen palette controls the appearance of line features in the view, offering various types of solid, dashed, or dotted lines, as well as the option of changing line weights. This is by far the simplest of the five main palettes, but similar design guidelines apply to its use.

Consider the geographical features represented by the line theme, and establish a visual hierarchy. For example, interstates can be represented by thick double lines; U.S. routes by thinner, single lines; and state and county roads as the thinnest lines. In a similar fashion, you can establish a visual hierarchy of rivers by representing those with the largest volumes with thick lines and their tributaries with thinner line weights.

Legibility is always a consideration with line work. Print out a few draft maps before setting line weights to ensure that your printer can show the smallest line weight used on the map, and that the largest line weight does not obscure other important map features.

> ✓ **TIP:** *When showing themes with a lot of lines, use lighter gray shades or lighter colors to "drop back" lines, rather than through the use of solid black or dark colors. This reduces the visual confusion of a lot of dark lines resembling a "ball of yarn."*

The default window for the Pen palette offers 28 basic line types. Below this window, there are drop-down menus for Size (in points), Cap, and Join. Size determines line thickness, and ArcView defaults to a near invisible tenth of a point. For most mapping purposes, you will use between .5 pt to 3 pt width lines, and can manually enter any size between, such as 2.5 pt, 2.6 pt, and so on.

Cap determines how ArcView draws the end of lines, either as Butt, Round, or Square, which are options found in the drop-down menu. Butt is the default, which terminates the line at right angles to its direction at the last vertex. Round adds a half circle at the last vertex, whereas the Square option squares the line off by half the line width beyond the final vertex.

Join determines how the line appears not at its ends but between vertices, as Miter, Round, or Bevel. A Miter join results in the line being sharp and angular at each vertex where the line changes direction. A Round or Bevel join results in a smooth, curved path at each vertex where the line changes direction. The Cap and Join menus are shown in the facing illustration.

The Cap and Join options are used infrequently, but are very useful in particular circumstances. An example is how roads are depicted on a map. If a theme containing road locations with coarse resolution has vertices representing actual control points on the road centerline, the road's line can appear quite jagged. Few roads actually jerk from one point on the landscape to another, and the resulting line on the map can be very misleading. You can control this feature by changing the Butt and Join options to Round, which will depict the road as a smoothly curving feature.

Pipelines, in contrast to roads, are used for utilities, and frequently do suddenly change direction as they pass underground from one building to another. In this case it would be inappropriate to change the Butt and Join options from their square defaults.

Cap and Join drop-down menus and their results.

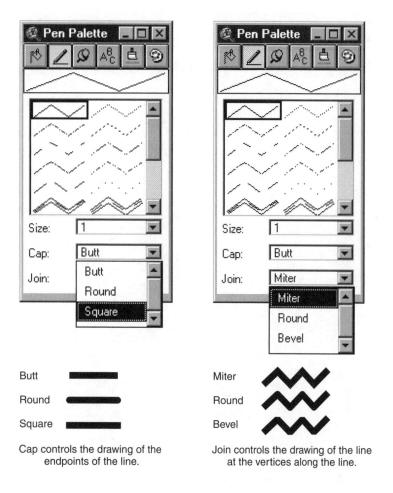

Cap controls the drawing of the endpoints of the line.

Join controls the drawing of the line at the vertices along the line.

✓ **TIP:** *Specialized lines and point markers used in the Geology palette are presented in a layout document that comes with ArcView. Navigate to the ArcView samples directory, which is normally found on Windows PCs in the* c:\ESRI\AV_GIS30\ARCVIEW\Samples\ other\aisym.apr *directory. The* aisym.apr *project has an attractively designed E-size layout with pen and marker symbols explained in detail for use in geological mapping.*

Using Symbolizer to Create Line, Polygon, and Pattern Symbols

Symbolizer is a free ArcView extension produced by Flat World Technologies and available on the web at *http://spatial-online.com*. Rather than endlessly searching the web for palette files that contain just the symbol you want for a mapping project, you can quickly produce what you want from Symbolizer.

Once Symbolizer, shown in the following illustration, is downloaded from the web and the extension file, *symbolizer.avx*, placed in the usual extensions folder found at *c:\esri\av_gis30\arcview\ext32*, you can start Arc-View, open a project, and activate Symbolizer from the Extensions dialog box found in the File menu. Symbolizer adds one button to the left of the Help button in both the view and the layout document windows. Clicking on either button gives you access to the Symbolizer controls, starting with line symbols.

Symbolizer extension showing line symbol options.

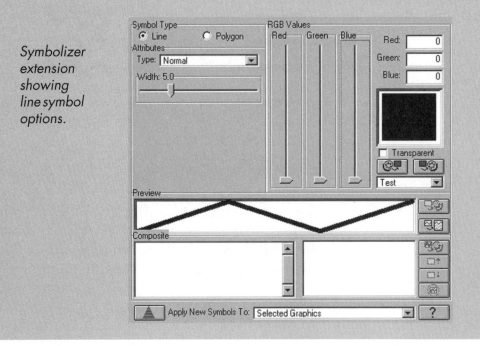

Symbolizer allows you to choose from 14 line types, such as dashed, dotted, and zigzag, and control the line width, pen size, and spacing of dashed lines. RGB colors can be assigned to the lines with the use of three color value sliders, or can be typed in as numeric values ranging from 0 to 255. A nifty advanced feature allows you to construct composite line symbols, which consist of multiple line symbol types laid on top of one another.

When you are finished constructing your line symbol, you can apply it to the active theme's legend, a graphic in the view or layout, or to the Pen palette in the Symbol Window. If applied to the Pen palette, you can use the Palette Manager to save the new palette as an ArcView palette file (an *.avp* file), which can be loaded into other ArcView projects.

Symbolizer also solves one of the ArcView problems of not being able to apply dashed or dotted lines to the outline of a polygon theme. Create a dashed or dotted line in Symbolizer, and apply it to the active theme's legend, making sure your active theme is a polygon. Previously without Symbolizer, you had to convert the polygon theme to a line theme in the view before dashed lines could be used.

With polygon features, Symbolizer allows you to create custom RGB colors for fills and lines (similar to the Color Picker extension). Symbolizer also offers the ability to create custom patterns by loading common raster files such as GIF or TIFF files and applying them to a pattern type. This is very handy for many types of specialized maps, such as geologic maps. Once the custom fill has been created, it can be applied to the active theme's legend, to a graphic, or to the Fill palette. Symbolizer, shown with polygon symbol options in the following illustration, is a great tool to add to your cartographic toolbox.

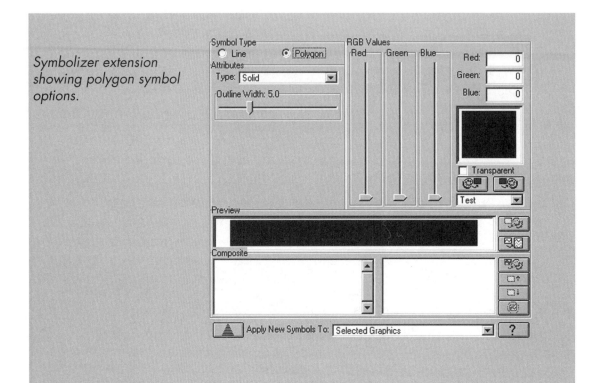

Symbolizer extension showing polygon symbol options.

Printing ArcView Palettes with the Symbol Dump Script

As previously mentioned, one of the continuing problems with map production is the pursuit of WYSIWYG, trying to get what you see on the computer screen to match what you get on the printed page. When you start to design a map, it is often difficult to determine what exact symbols or colors to use, and much time can be wasted revising colors and changing symbols after you discover they do not print clearly and legibly on the final copy.

One way to get around this problem is to print entire ArcView symbol palettes before you use them on the map, and then choose the symbols or colors that look best for your purposes from the printed copy. To do this, you need a handy Avenue script called *symdump.ave*, available from the ArcScripts page on the web at *http://www.esri.com/arcscripts*.

Download *symdump.ave* from the Web to your hard drive. Unlike Arc-View extensions, Avenue scripts do not need to be put in any particular directory on your hard drive. However, it is a good idea to create an *avscripts* directory to which you can save useful script tools. Open an Arc-View project, and add any new palette files you are interested in using with the Palette Manager's Load option. To print the default palettes, do not load anything.

In the ArcView Project Window, click on Scripts and select New. A blank script window appears. Go to the Script menu and go down to Load Text File. Use the dialog box to find the location of *symdump.ave*, and click on OK. The Avenue code making up the script should appear in the window. Go to the Script menu and drop down to Properties. In the Properties box, change the name of the script from Script 1 to *Layout.Symbol-Dump*.

Next, go up to the Script button bar and click on the Compile button. Now you are ready to run "symbol dump." Go to the button bar and click on the Run button.

After the Run button is pressed, a dialog box will prompt you to choose which of the Symbol Window palettes (Color, Fill, Marker, Pen, or Text) you want to put in a layout. Each symbol or color will be placed as an individual graphic in a layout. Complicated palettes may result in 5 to 20 or more layout documents. You can now print the palettes you are interested in from the layout window, and decide which symbols or colors would work best in your map.

Summary

The Symbol Window palettes provide a bewildering number of options for symbolizing point, line, and polygon data. There is no one "right way" of symbolizing each type of data, and this is the place to creatively experiment with different types of symbolization. You need to consider what purpose the map will serve and strive to establish a clean, legible presentation with a visual hierarchy of symbols that is easily understandable by your audience.

Chapter 6

Typography in the View

0 ____ Miles ___ 500

The act of placing lettering on a map remains one of the most challenging tasks facing a cartographer. The art and science of lettering, called typography, has come a long way in the digital age, but placing words on the map can still be a very tedious chore.

Two decades ago, type was placed on a map using press-on lettering or by placing on the map small pieces of wax-backed paper containing entire words produced on a photo typesetter. All of this changed in 1984 with the invention by Adobe Systems of digital type and electronic fonts for use on the computer. The invention of digital type started a renaissance in the discipline of typography, allowing type designers the freedom to create thousands of type styles and the ability to resurrect old classics for use in the computer age.

Text is added to a map for three general purposes: to accurately label map features in the view, to document map themes in the legend, and to populate margin information, such as the map title and source statement. By far the most involved task is that of labeling geographic features on the map itself, which can range from a mere handful of names to thousands of labeled map features. Clearly, this is no task for the faint of heart!

Digital typographic tools have made the cartographer's work much easier, but labeling map features with type has been one of the weakest points of the ArcView software. In fact, you need to closely consider how much labeling each map product demands. ArcView projects with extensive labeling tend to bloat up to enormous file sizes and become difficult to manage and work with. Maps with thousands of labels are better finished in other graphics software packages, such as Adobe Illustrator or Macromedia Freehand, which have extensive typographic tools and font management options.

Recent improvements to ArcView have helped, such as the introduction of new labeling tools in ArcView 3.1 and the creation of ArcView extensions that help control the text labeling process. However, before examining these tools, you need to understand some type terminology and general conventions for the use of type on a map.

Type Terminology

More people than ever use word processors, page layout programs, and graphic applications on computers, but few people take the time to master digital typography. Here is an overview of type terminology that will prove helpful when it comes time to put text on a map.

Typography is the science (and art) of creating, managing, and placing type in documents, which today is almost entirely digital (that is, managed on a computer). Lettering is a traditional term used to describe the process of placing letters and words (type) on a map. Feature labeling is the process of lettering map features with the appropriate names. This process takes place in the map display portion of the view within ArcView.

Type consists of characters, typefaces, fonts, and font families. A character is the individual letter of the alphabet, number, or symbol that makes up a typeface. A typeface is the design of the characters that along with several other factors, such as type weight and pitch, make up a font. Common digital typefaces (shown in the following illustration) include Arial, Helvetica, and Times. Typefaces and fonts come in two basic styles, serif typestyles and sans serif typestyles.

Serif Type

Times New Roman 10pt:

Labeling maps is a lot of work — Regular —

Labeling maps is a lot of work — Italic —

Labeling maps is a lot of work — Bold —

Labeling maps is a lot of work — Bold Italic —

Adobe Garamond 10pt:

Labeling maps is a lot of work — Regular —

Labeling maps is a lot of work — Italic —

Labeling maps is a lot of work — Bold —

Labeling maps is a lot of work — Bold Italic —

Sans Serif Type

Arial 10pt:

Labeling maps is a lot of work

Labeling maps is a lot of work

Labeling maps is a lot of work

Labeling maps is a lot of work

Adobe Helevetica 10pt:

Labeling maps is a lot of work

Labeling maps is a lot of work

Labeling maps is a lot of work

Labeling maps is a lot of work

Type styles.

Serif typestyles have small decorative traces at the tips of their characters. Although this makes the characters more complicated, it also makes them easier to read (less tiring to the eye) when text is placed in large blocks or on entire pages of a book. Serif typestyles such as

Times Roman are often used to label hydrographic features, such as rivers and lakes.

Sans serif typestyles consist of simple, clean characters without serifs, which make individual words set in sans serif easier to read at small point sizes (below an 8-pt type size). Conversely, serif typestyles are more difficult to read (and very tiring to the eye) when set in large blocks of type on the printed page. Sans serif typestyles (such as Arial or Helvetica) are commonly used for labeling terrestrial features on maps.

Fonts are the names that appear in the scrollable list in the Symbol Window's Font palette. Fonts with similar designs are organized into font families. One of the font families available in the Font palette is the Arial family of fonts, including Arial (regular), Arial Black, and Arial Narrow.

> ✓ **TIP:** *Cartographic designers will choose two, or perhaps three, font families to use throughout a large mapping project. In this manner, a reference map with thousands of pieces of type will appear to have a consistent and easily recognizable design.*

Types of Type

There are three main types of fonts in use on computers today, which reflect developments in digital typography over the last 30 years. The original means of representing type characters on the computer screen used pixels or bitmaps to construct bitmapped fonts. Bitmapped fonts are easy to create and display, but appear stair-stepped (jagged) when enlarged on the computer screen or on paper. Some bitmapped fonts are still found on computers today, but should not be used to produce professional-looking maps.

PostScript fonts use the PostScript page description language developed by Adobe Systems in 1985. PostScript allows the creation of resolution-independent outline fonts that appear the same at any level of enlargement on the computer screen or on paper. PostScript fonts remain the standard within the publishing industry, and cartographers use them for high-quality atlases and map books.

TrueType fonts were invented in 1991 by Apple Computer and Microsoft as a lower-cost alternative to PostScript fonts. Today, True-Type fonts are probably the most widely used fonts, and work well for most in-house mapping purposes. Problems remain with TrueType fonts when sent to high-end image setters in the map publishing process. You can tell which fonts are which type by using a utility such as Adobe Type Manager to load and organize fonts on your computer.

Organizing Fonts with Adobe Type Manager

Adobe Type Manager is available from Adobe Systems of Mountain View, California *(http://www.adobe.com)*. Type Manager comes in two forms, basic and deluxe. The basic Type Manager usually ships with any Adobe software, but Type Manager Deluxe must be purchased separately.

Type Manager allows you to view all fonts on your computer (both Post-Script and TrueType), and to print samples of them and organize them into sets for use with particular projects. If your organization has a particular "style sheet" for publications or maps, you can create a set with just those fonts. The first of the following illustrations shows Adobe Type Manager Deluxe's master font list. The second of the following illustrations shows fonts organized as sets in Adobe Type Manager Deluxe.

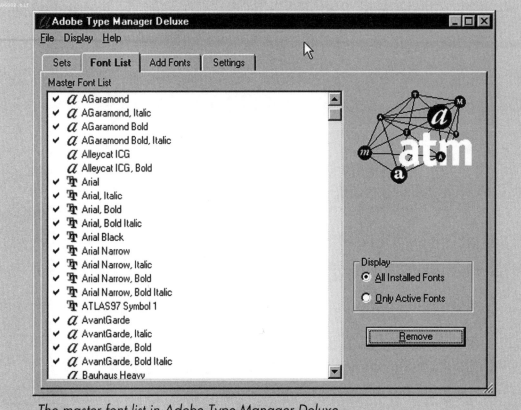

The master font list in Adobe Type Manager Deluxe.

Adobe Type Manager Deluxe

File Set Help

| Sets | Font List | **Add Fonts** | Settings |

Destination: Regular Type

Source: Browse for Fonts

C:\PSFONTS\PFM

 Pfm

☐ 🅐 [New Set]
☐ 🅐 Decorative Type
☑ 🅐 Default Type
☐ 📁 Regular Type
　　☐ 𝒜 Benguiat
　　☐ 𝒜 Benguiat, Bold
　　☐ 𝒜 Benguiat Gothic Book
　　☐ 𝒜 Benguiat Gothic Book, Italic
　　☐ 𝒜 Benguiat Gothic Heavy
　　☐ 𝒜 Benguiat Gothic Heavy, Italic
　　☐ 𝒜 Benguiat Gothic Medium
　　☐ 𝒜 Benguiat Gothic Medium, Italic
　　☐ 𝒜 Benguiat Gothic Medium, Bold
　　☐ 𝒜 Benguiat Gothic Medium, Bold Ital
　　☐ 𝒜 Bernhard Modern Roman
　　☐ 𝕋 Bookman Old Style
　　☐ 𝕋 Bookman Old Style, Italic
　　☐ 𝕋 Bookman Old Style, Bold
　　☐ 𝕋 Bookman Old Style, Bold Italic

𝑎 Bookman, Bold
𝑎 Bookman, Bold Italic
𝑎 Bookman Medium
𝑎 Bookman Medium, Italic
𝑎 Bookman Medium, Bold
𝑎 Bookman Medium, Bold Italic
𝑎 Brush Script, Italic
𝑎 CaflischScript Regular, Italic
𝑎 CaflischScript Regular, Bold Italic
𝑎 Carta
𝑎 Caslon 224 Black
𝑎 Caslon 224 Black, Italic
𝑎 Caslon 224 Book
𝑎 Caslon 224 Book, Italic
𝑎 Caslon 224 Medium

☐ Add without copying files

| Remove | Add |

Fonts organized as sets in Adobe Type Manager Deluxe.

Labeling Map Features — The Uses of Text

Map features should not be labeled haphazardly. There is always a reason to label some map features in one manner and other map features in a different manner. It is good to review the purpose of placing text on maps before beginning the often tedious task of labeling.

Text on a map can be used in four ways: to name a feature, to locate a feature, to indicate a type of feature, and to indicate size. These uses have equivalent names as text labeling systems: literal, locative, nominal, and ordinal. The following illustration shows text labeling systems.

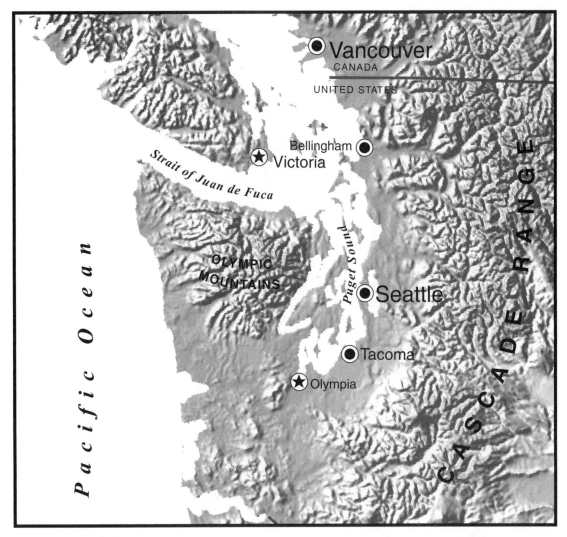

Text labeling systems.

Naming a feature is the most obvious use of text. You name the point features that represent cities to indicate, for example, that San Francisco is different from San Jose, and that both are different from Oakland when making a map of the Bay Area in California.

Text can be used as a locative device as well. For example, if you are using polygons to represent cities, rather than points, the words _San Francisco_ can be placed entirely within the polygon representing the city limits. In a similar fashion, when you are labeling mountain ranges on a physical features map, the words _Sierra Nevada,_ for example, can be curved and spread to roughly overlay the full extent of the actual mountain range.

Different styles of text can be used to differentiate between different nominal classes of map features. A typical example of this is the use of italic type in labeling hydrographic features such as rivers and lakes. In this case, the italic text is frequently the same blue color as the water body it is associated with, making it easier for the map reader to quickly read and understand the labeling system.

Care must be taken when coloring text. Most type on maps is black because black stands out well when used in small characters on a white background. Black text on white background has good figure-to-ground contrast, and makes type easy to read for the map user. Other colors, including blue, make text more difficult to read, especially if used on a background darker than white.

> ✓ **TIP:** _To clean up busy maps that contain a lot of text labels, color text that is less important to the map's purpose a shade of gray, rather than black. When showing the location of cities in the western United States, for example, keep the city names in black, but drop back the state names to a medium to light gray._

> ↝ **NOTE:** _Terminology is important here. Traditionally, the term over-posting was used to refer to one piece of text being placed over another_

piece of text, or obscuring a map feature. ArcView uses the term over-lap *to refer to the same situations.*

In order to make this nominal labeling system work, it is important to apply it consistently throughout the map. If you are using the Times Roman Italics typestyle for some river features, you should use it for all river features on the map.

Different sizes of text can also be used to denote different sizes of map features. Cities with larger population sizes will have text with larger point sizes, and cities with smaller populations will have smaller text labels. Combined with larger symbols for larger cities, the technique of using ordinal text reinforces the city size factor with the map reader.

Labeling Map Features — Text Positioning Guidelines

Once you have identified the map features you want to label, and have considered possible text labeling systems, you are ready to start placing type on the page. Exactly where do you place it? The following are general conventions that can be used as guidelines for text placement. By adhering to these conventions, you make the map easier to read and understand.

Guideline 1: A Label should be on either land or on water, but not on both.

This is an easy convention to adhere to when mapping along an oceanfront, but becomes more difficult to follow when dealing with smaller lakes and rivers. Work to include the name of a lake within the lake's shorelines. If this cannot be done, place the name to the right or left of the lake.

It is important to retain a consistent type size while labeling these features. Use a larger type size for big lakes, and smaller size for the rest of the water bodies. Do not use a different type size for each lake just to fit the name within a lake. An example of hydrographic labeling is shown in the following illustration.

Labeling hydrographic features.

Guideline 2: Break a geographic feature behind a piece of type rather than the converse.

This helps readability, but requires more work. There are three methods that help out in this case. The first is to make the geographic data a lighter gray or a lighter color so that the black text will stand out when printed over the top of it. This is only possible if lightening up the geographic data does not detract from the purpose of the map.

The next method, shown in the following illustration, is to create a separate theme with features or graphics that act as "blockers" or "knockouts" when placed between the text and the theme containing features the text overprints. The features used as the "knockouts" should be the same color or shade as the map background.

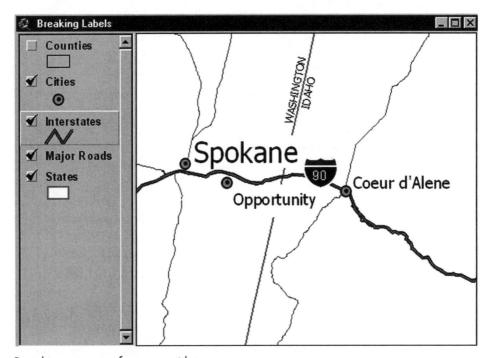

Breaking up map features with type.

Sometimes, neither of these methods is helpful, and you need to actually split map features in a theme to have correct type position. Never split up the original version of a theme, because this process will make the theme unusable for geographic analysis. Instead, make a real copy of the theme and then start editing, using the line or polygon splitting tools in the Draw Tools drop-down palette to split the feature.

   ✓ **TIP:** *The following are three methods of copying your geographic shape-file data.*

     • *Use the Convert to Shapefile command in the view's Theme menu.*

     • *Use the Copy command in the Shapefile Manager, found under Manage Data Sources in the view's File menu.*

     • *Outside ArcView, in Windows Explorer, select the .shp, .shx, and .dbf files of a single shapefile and copy and paste to new files.*

   ↦ **NOTE:** *Remember, using Copy Themes in the view's Edit menu only creates another link to a geographic data set and does not actually copy the original data. If you start to edit a theme copied in this manner, you will corrupt the original data.*

Guideline 3: Use curved and angled type sparingly.

Anytime type is curved or angled it becomes more difficult to read. Curved or "splined" type is appropriate when applied to the labeling of rivers or mountain ranges, and can add an attractive touch to any map when done carefully. Always used curved text for such situations and not angled text.

Use steeply angled type rarely, mainly to label point symbols you want to stand out in some fashion. When curving or angling a label, never label text upside-down. Examples of curved labels are shown in the following illustration.

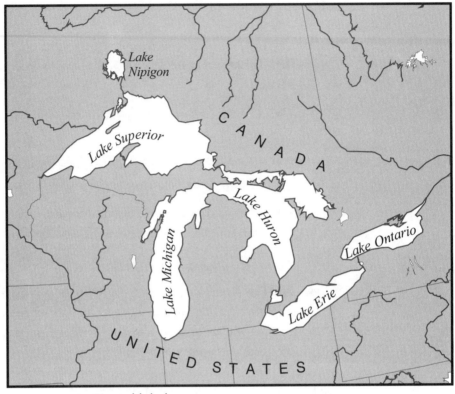

Curved labels.

Guideline 4: Match the type orientation to the orientation structure of the map.

The orientation structure of a map is traditionally associated with its scale. Small-scale maps (1:100,000 scale and up) frequently have latitude and longitude lines that act as a reference grid for map features. Match the orientation of type (by slightly angling the label) to the longitude lines when working with small-scale maps. The precise angle of a piece of type on such a map will depend on where the type is placed in relation to a parallel line of longitude. On large-scale maps (below 1:100,00 scale), orient the type for most map features as horizontal (parallel) to the top and bottom edges of the map page.

Guideline 5: Spread the characters within a piece of type as little as possible.

Most type used on the map should not be spread at all. Spreading type horizontally (kerning), as shown in the following instration, is a good technique when labeling a large feature that remains in the background, such as a political jurisdiction. When labeling curving features such as rivers, take care not to spread type too much, which makes a label too spread out to be read as one word.

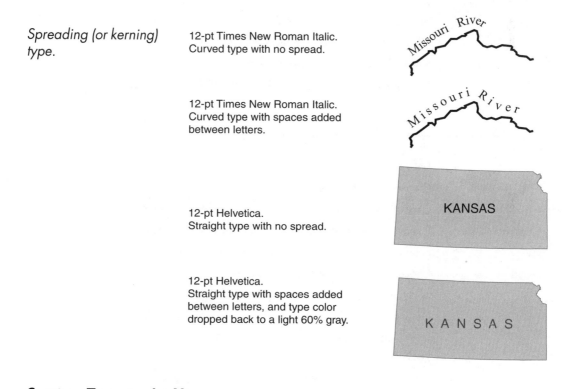

Spreading (or kerning) type.

12-pt Times New Roman Italic. Curved type with no spread.

12-pt Times New Roman Italic. Curved type with spaces added between letters.

12-pt Helvetica. Straight type with no spread.

12-pt Helvetica. Straight type with spaces added between letters, and type color dropped back to a light 60% gray.

Setting Type in ArcView

As mentioned previously, text is added to a map for three general purposes: to label map features in the view, to document map themes in the legend, and to populate margin information. These purposes are achieved by using a variety of tools in ArcView's view and layout documents. In general, the process to follow when adding text to a map is to accurately label the Legend first, identify what map features are important to the map's purpose and label them in the view's map display, and add type for the map title and margin information last. Table 6-1, which follows, presents a flow chart for this process.

Table 6-1: Flow Chart for Labeling an ArcView Map

Location and Tool	Task
Planning	How much labeling is needed? • Just a few labels: Use ArcView Text tool • More labels: Use ArcView Label tool • Hundreds of labels: Use Auto-Label or Multi-Labeler; consider buying AVALabel • Thousands of labels: Consider finishing map in Dedicated graphics program
In the View	Use Theme Properties and the Legend Editor to label TOC. Determine which themes need labels. Create map title in View Properties box.
In the Table	Check the attribute field you want to use as feature labels. If necessary, edit field in the table with Field Calculator to create labels.
In the View	Set a reference scale for labeling based on final map size. Choose a type style and size for labels. Set type style and size in Text and Label Defaults box. Use the Text, Label, or Auto-Label tools to label map features. Use curved (splined) text for rivers, mountain ranges, etc. Move labels to get precise position and avoid overlaps. Attach text to individual themes with Attach Graphics.
In the Layout	Place map and legend using View and Legend Frame tools. Add marginal text, source statement, and title with Text tool. If necessary, copy and paste blocks or text from a word processor in the Text tool dialog box.
On Paper	Print map and check labels for readability, position, and spelling.

Creating Text for the Map Legend

Text in a map legend is generated from the Table of Contents in the view. Whatever themes are checked on as visible in the TOC will appear listed in the legend in the layout. The process for creating the legend text involves the following four primary steps, during which you can change text or type style.

1. Renaming themes on the Theme Names line in the Theme Properties box

2. Naming the Values in a theme's legend in the Label column in the Legend Editor

3. Adjusting type styles in the Table of Contents Style Settings

4. Fine tuning the Legend in the layout document

Renaming Themes

After adding themes to a view, it is always a good idea to rename them to make them more understandable. (See the sidebar "Problems of the 8.3 Naming Convention" in Chapter 3.) Click on the theme's name in the TOC to make it active, and go to Properties under the Theme menu. The Theme Properties dialog box, shown in the following illustration, appears.

Theme Properties dialog box.

The Theme Name line allows you to rename the geographic data in the TOC without changing the original data set's name. Choose a name that is quickly understandable by your intended map audience. The name can be several words long, but it is best to keep it short so that it can all appear on one line in the TOC.

A good example is a point theme representing cities in California, which as a shapefile might appear as *cacities.shp*. If the theme represents just the largest cities in the state, you could rename it "Major Cities" or "Major Cities in California." Frequently, cities will be symbolized by population size; therefore, in this case a good theme name might be "Cities of 1 million people or greater." If the theme is a comprehensive list of all cities in the state, a better name would be "Cities and Towns." Any of these theme names quickly conveys much more information than the default name *cacities.shp* or the overly simple "Cities."

Individual Theme Legend Text

Once a theme has been symbolized in the Legend Editor, shown in the following illustration, a legend description appears beneath the theme's name in the TOC. This description can be hidden by using the Hide/Show Legend option under the Theme menu. The Legend Editor, as a default, creates the legend labels directly from the values (which can be numbers or text) used to classify the theme. These can be changed in the Label column within the Legend Editor.

As with theme names, the legend labels should be easily understandable to the map's intended audience. Often default numeric legends can be confusing to general audiences. A legend showing the amount of a product sold in each city in a state might be meaningful to the marketing department of a company, but appear confusing or useless when shown to the stockholders in an annual report. For a non-technical audience, a legend showing High, Average, and Low sales areas would be much more useful.

Labeling in the Legend Editor.

TOC Style

The text in the view's table of contents defaults to a standard Arial 10-pt font on a PC. Starting with ArcView 3.1, this can be changed, along with line symbol styles for the legend, in the TOC Style dialog available from the View menu. The TOC Style dialog box includes the ability to use any font, font style (bold, italic, and so on) and type size available in your system as a global setting for all themes in the TOC. Table of Contents style properties are shown in the following illustration.

Table of Contents style properties.

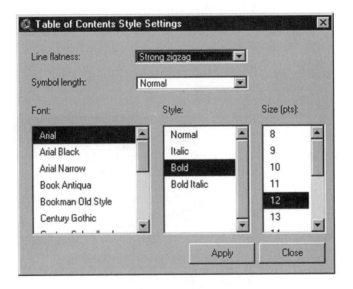

The degree of zigzag for line themes, and the length of symbol for line and polygon themes, can be selected in the TOC Style dialog box. Zigzag is set with Line Flatness, having the option of strong zig-zag, moderate zigzag, or flat lines. The classical zigzag line in the legend has long been a trademark of maps produced in ArcView. Flat lines are a great alternative, and present a much cleaner appearance.

Symbol length can be set as Normal, Short, Long, and Very Long. The longer symbol length gives more of the symbol for the map reader to compare to the map features. Setting a longer symbol length is helpful when using dashed or dotted lines, or when filling polygons with light colors or complex patterns.

A larger symbol size, especially with polygons, can be quite helpful in overcoming a perception problem with legends. When a small area filled with a specific color (in the legend) is compared to the same color in a larger area (in the map), the smaller area's color is perceived as darker by the map reader. If the map has dozens of colors (common with land cover, vegetation, soil, and geologic maps), it can be very difficult to match the legend color to the map feature color. Combine a larger symbol length in the legend with map features labeled with a short piece of text to act as a key to make complex color maps easier to understand.

Options set in TOC Style do not affect type or symbols in the map display of the view window, but only the text and symbols in the table of contents. Settings in TOC Style do carry over to the Legend when that is created from the Legend Frame tool in the ArcView layout document.

Legends in the ArcView Layout

There is always the option of changing the appearance of your map legend at the last moment in the ArcView layout document. For more information, read about creating map legends with the Legend Frame tools in Chapter 8.

Labeling Map Features

There are three methods of labeling map features in the basic ArcView package: using the Text tool, using the Label tool, and using the Auto-label option. You can use one tool to label the map, or all of the tools in combination, employing each tool's particular strengths. Going beyond basic ArcView, Map Logic Corporation offers a fine

AV extension called ArcView Advanced Label (AVALabel 2.3) that refines and automates much of the tedious chore of map feature labeling. The sections that follow examine all of these tools, starting with the simplest and proceeding to the most involved.

The Label Tool and the Text Tool

The first four tools of the Label and Text Tool drop-down menus (shown in the following illustration) are nearly identical, but have one distinct difference. The Text tool requires that you type in the proper feature name, whereas the Label tool uses type from a field in a theme's attribute table.

The Text tool in the view or in the layout can be used to quickly label important points in the map, if you know what you want to name. This tool should only be used for small amounts of type. You are responsible for creating the correct names for a map feature because the Text tool does not generate its labels from a theme's attribute table.

Text and Label drop-down menus.

The Label Tool Drop-down Menu

- Label Tool
- Callout Label
- Bullet Leader Label
- Banner Label
- U.S. Interstate Label
- U.S. Route Label
- Generic State Highway Label
- Generic Square Highway Label
- Generic Oval Highway Label

The Text Tool Drop-down Menu

- Text Tool
- Callout Text
- Bullet Leader Text
- Banner Text
- Drop Shadow Text
- Spline (Curved) Text

To use the Text tool in the view, first go to the Window menu or to Show Symbol Window, or use the keyboard command <Control +P>. The Fill palette will appear as the default. Click on the Text button to bring up the Font palette. Choose a font, and a type size and type style.

✓ **TIP:** *This process of choosing a type size and type style before using a "type" tool in ArcView ensures that the result will be the text size and style you want, not the Arial 14-pt font that is frequently the ArcView default. Otherwise, you must choose the pieces of text you have already placed with the Selection tool, and then use the Font palette to change type styles.*

To change the default for the Text tool or any other labeling tool, use the Text and Label Defaults dialog box (shown in the following illustration) from the Graphics menu in the view. Here you can change the default type size, type style, and color for any text object, and set the outline and fill properties for highway shields and callout boxes.

Text and Label Defaults dialog box.

There are six tools available in the Text tool drop-down menu, and each tool places type in the map display in a slightly different manner. The basic Text tool places text horizontally next to the map feature you want to label and brings up the Text Properties dialog box. In the dialog box, you enter text in the text window, making sure to press the <Enter> key when you want to stack the text on another line. You have four other options in the Text Properties box: Horizontal Alignment, Vertical Spacing, Rotation Angle, and Scale Text with View.

Horizontal Alignment allows you to left justify, center, or right justify text. Vertical Spacing controls the amount of space between stacked text based on line height measured from 1.0 to 2.0 (which is related to type size in points). Rotation Angle will rotate a piece of text from 0 to 360 degrees, with the default being 0 degrees or horizontal. The Scale Text with View option, when checked, allows text to get larger in size as you zoom into a view, and get smaller in size as you zoom out. If this option is unchecked, text will remain the same point size no matter how much you zoom in or out of a view.

The second text tool in the drop-down menu creates Callout Text. After clicking once on the tool, you must click on the map feature you want to label and then drag a small distance away to place the callout box. The Text Properties dialog box appears, prompting you to enter your text. This tool can be very handy when you want to draw the reader's attention to a few features of importance on the map, but has the disadvantage of creating opaque boxes that can cover significant parts of the map.

The default style of the callout box includes the pointer and the box, which is surrounded by a gray drop shadow and filled with a light fill color. To change these colors globally, use the Text and Label Defaults option under the Graphics menu. To change just one callout box, go to the Color palette in the Symbol Window and use the Foreground option to change the box fill, and the Background option to change the color of the drop shadow.

With the callout box selected, use Ungroup in the Graphics menu and the Vertex Edit tool from the View Window's tool bar to change the position of the callout box, the pointer, and the drop shadow. This is helpful when you need precise placement of several callout boxes but do not want to obscure important map features.

The Bullet Leader Text tool is similar to the Callout Text tool, but uses a dot or bullet to mark the location of the map feature, and a line that leads to the text identifying the feature. In the same manner as the Callout Box text tool, the Bullet Leader Text tool allows you to click on a map feature and drag to the location you want for the text placement. There is no box, pointer, or drop shadow to cover underlying parts of the map. However, you may end up with many black leader lines, which can be confusing on a map that has a lot of black line work, such as a road or utility map.

The Banner Text tool creates a box with a drop shadow over the map feature you want to name. Banner Text is great for drawing attention to a few features, but does obscure significant portions of the map. Use this tool sparingly.

The Drop Shadow Text tool places a black piece of text with a dark gray shadow behind, below, and to the right of the text. This is a nice effect for a few larger text labels, but can prove difficult to read if used for many features.

Much more useful is the Spline Text tool, which places text along a curved line. After selecting the tool from the Text Tool drop-down menu, you begin placing a curved line by single clicking on the map display and double clicking to end the line. The Text Properties dialog box appears, allowing you to name the map feature. When done with care, the Spline Text tool can add a lot of visual value to the text labels on your map.

> ✓ **TIP:** *When labeling curved line features, such as rivers, with the Spline Text tool, always offset the curved text line a short distance from the feature you want to label, in order to increase readability. When labeling curved polygons or relief features, such as mountain ranges, place the curved text line near the center of the feature, and extend the line the full extent of the feature.*

Annotation Themes

You can use already created labels called annotation themes produced in ArcInfo or AutoCAD. Annotation themes, shown in the following illustration, contain just the label for a feature, and not the feature itself, and you are relying on the producer of those themes to label the features you want to present on your ArcView map. The text label cannot be edited in ArcView, but can be changed in its native ArcInfo or AutoCAD. Annotation is added just like any other geographic theme, by using the Add Theme (plus) button in the view's button bar.

Annotation is represented, oddly enough, as a polygon theme with "text blocks" when added to a view. The label inherits certain unchangeable characteristics from ArcInfo. These characteristics include the exact placement of the label in relation to the map feature, the rotation angle (if any) of the label, and the text itself. Although none of these can be changed in ArcView, the type style and color can be changed in the Legend Editor.

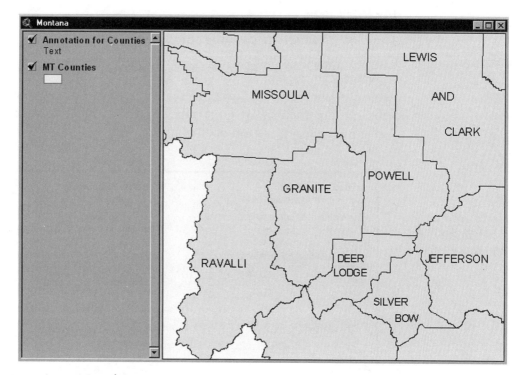

Annotation themes.

Whether you can change the type size in ArcView depends on how the annotation coverage was saved in ArcInfo. If the annotation coverage was saved in page units in ArcInfo, the type size can be changed in ArcView, by using the Legend Editor and the Font palette. If the annotation coverage was saved in map units in ArcInfo, the type size cannot be altered.

Annotation added from CAD themes have the text size "hardwired in" as measured in map units. You can change the text color and type style in the Legend Editor, but not the size of the font. With these limitations, annotation themes are only as useful as their original creators have made them, and are not commonly encountered in the ArcView world.

The Label Tools

The most common means of feature labeling is using text from a field in a theme's attribute table, as shown in the following illustration. Here you can use the label tools in the view's toolbar, or one of the two Auto-label features. This process assumes that the text you want is found in the theme's attribute table, and that the producer of the theme has done a good job of attribution.

Fields used as labels in the Theme Attribute table.

A Poor Label Field in a Theme Attribute Table	**A Better Label Field in a Theme Attribute Table**
The Name field containing County Names is in all caps with underscores between words.	The Name field containing County Names is in upper and lower case with spaces between words.

Attributes of County.shp		Attributes of Counties.shp	
Shape	*Name*	*Shape*	*Name*
Polygon	GOLDEN_VALLEY	Polygon	Golden Valley
Polygon	GRANITE	Polygon	Granite
Polygon	HILL	Polygon	Hill
Polygon	JEFFERSON	Polygon	Jefferson
Polygon	JUDITH_BASIN	Polygon	Judith Basin
Polygon	LAKE	Polygon	Lake
Polygon	LEWIS_CLARK	Polygon	Lewis and Clark
Polygon	LIBERTY	Polygon	Liberty
Polygon	LINCOLN	Polygon	Lincoln
Polygon	MADISON	Polygon	Madison
Polygon	MCCONE	Polygon	McCone
Polygon	MEAGHER	Polygon	Meagher
Polygon	MINERAL	Polygon	Mineral
Polygon	MISSOULA	Polygon	Missoula
Polygon	MUSSELSHELL	Polygon	Musselshell
Polygon	PARK	Polygon	Park
Polygon	PETROLEUM	Polygon	Petroleum
Polygon	PHILLIPS	Polygon	Phillips
Polygon	PONDERA	Polygon	Pondera
Polygon	POWDER_RIVER	Polygon	Powder River
Polygon	POWELL	Polygon	Powell
Polygon	PRAIRIE	Polygon	Prairie

Before any of the eight Label tools can be used, you must choose what field to use from the theme's attribute table. To do this, with your selected theme active in the TOC, go to Properties in the

Theme menu, and select Text Labels. On the Label Field line in the Text Labels properties box, choose a field from the theme's attribute table. Next, choose a text position, relative to the map feature. An example of text label properties and positions is shown in the following illustration.

Text label properties and positions.

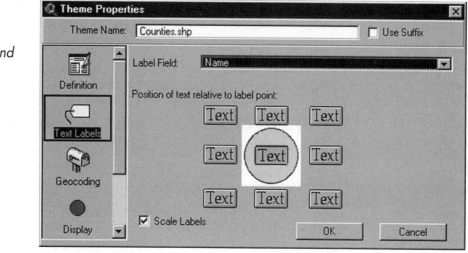

> ✓ **TIP:** *A quick reference for text position: For most point features, choose right or left. For line features such as rivers, choose above. For line features such as roads, choose center. For polygon features, use the center position as well. After the text is placed, you can always move it using the Pointer tool from the view toolbar.*

At the bottom left corner of the Text Labels properties box is the Scale Labels check box. The default selection is Scale Labels checked on, which means that text appears larger as you zoom in and smaller as you zoom out of a view window. This raises an important question when labeling a theme: At what scale should you begin labeling? A good choice is at the scale (the actual size) your map will be read by the map user.

✗ **WARNING:** *Choose your map reference scale carefully and stick with it! If you zoom in or out of the reference scale and continue to set type at the same type size, all of your labels will appear as different sizes when the map is printed!*

In regard to the previous Warning, this happens because ArcView maintains a default type size for labeling map features (usually 14 pt if you do not change it), no matter what scale you have zoomed in or out to. For example, say you are creating a letter-size map of major tourist attractions in Florida for a travel agency. You begin labeling, zoomed out to the entire state in the view (including Disney World and Everglades National Park), with the Label tool with 10-pt type. Then you zoom in to label a few sites in the Miami area.

When you zoom out to the state again, the 10-pt type for tourist sites around Miami will appear microscopic, and probably unreadable. In this case, however, if you label all map features while zoomed out to the entire state of Florida, your labels will all be readable when the map is printed at letter size.

➻ **NOTE:** *This serious problem can be avoided by using the Fixed Scale labeling tools within the AVALabel extension.*

Frequently, all of the information you need to label a feature may not be available in one attribute field, such as labeling cities with elevations above sea level. In this case, you may want to label the same theme twice by using "City Name" in the Label Field in the Theme Properties dialog box, labeling the cities, and then switching the Label Field to "Elevation" and labeling the cities a second time. This results in two separate pieces of text per city, rather than "Elevation" appearing in the same text block as "City Name."

It is smart to label everything you want on one theme in one work session, and then go on to the next theme that needs labels. After you have finished labeling a single theme, turn off all other themes in the view, and with the theme still active, go to Select All Graphics under the Edit menu. All of the text blocks you have placed will be selected with their black graphic handles visible at the four corners of each graphic block. Then use the Attach Graphics option under the Graphics menu to associate the text graphics with the original theme. In this manner, you can maintain a visual hierarchy of labels associated with each theme that will appear or disappear in the view if that theme is turned on or off.

If the Attach Graphics option is not used, text tends to "float" in the view and may not be turned off when themes are turned off. This can lead to confusion and poor label organization, especially when several themes have been labeled with dozens of pieces of text.

Once a label has been placed with any of the Label tools, the piece of text becomes a graphic object in the view. You can select the graphic text object with the Pointer tool by single clicking on the text and dragging it for more precise placement. Double clicking on the graphic object allows you to edit the text piece by bringing up the Text Properties dialog box, shown in the following illustration.

Text Properties dialog box.

Text Properties

Rosebud County

Horizontal Alignment:

Vertical Spacing: ▼ 1.0 lines

Rotation Angle: 0 degrees

☐ Scale Text with View

OK   Cancel

In addition, once the label has been placed with any of the Label tools, the resulting graphic text object is no longer live-linked to the theme's attribute table. For example, say you are placing labels for many city names on a map with the Label tool, and you notice several misspellings. Because you might use the city theme again, you edit the "city name" field in the theme's attribute table and correct the misspellings. The labels you have already placed in the view will not automatically change to the corrected spellings, and will have to be deleted and re-labeled.

When using names from a theme's attribute table, you should first review the name field and make sure that names are entered the way you want to portray them on the map. In many older GIS databases, the name "San Francisco" may be entered as "San_Francisco" or "SAN_FRANCISCO." Underscores between words are never used and words in all caps are infrequently seen on maps.

✓ **TIP:** *If you need to change text from all caps or remove underscores between words in a field in an attribute table, use the Field Calculator in the Tables document. This powerful tool can save you the enormous amounts of time it would take to retype text as upper- and lower-case characters.*

Start editing the table, create a new text string field from the Edit menu, and use the Avenue requests in the Field Calculator to populate the new field with text the way you want it to appear on the map in the view. See "Calculating a field's values" and "Performing operations on string fields" in the ArcView Help files for more complete information on how to accomplish these tasks.

Highway Symbol Label Tools

Last on the Label tool drop-down toolbox are the five highway sign label tools. These tools provide standard signs for U.S. Interstates, U.S. Routes, generic state highways, and generic square and oval highway shields. If a theme has a field in its attribute table with highway identification numbers in it, you can use these label tools to attractively label a road theme. Highway symbols are shown in the following illustration.

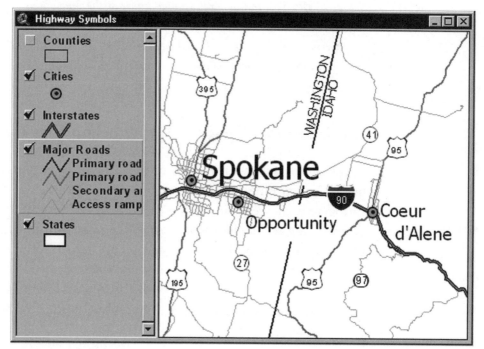

Highway symbols.

As with labeling any other theme, you must first set the Label field in the Theme Properties dialog box under Text Labels before choosing a highway symbol label tool. After clicking once on a tool, click once on a highway you want to label in the map display. The highway number automatically fills the highway symbol in the view. This assumes, of course, that you know enough about the data in the theme to correctly label a U.S. Interstate line with the U.S. Interstate Label tool, and not some other highway symbol.

If, as is common with most road databases, different types of roads are all included in one theme and you are unfamiliar with the data, use the Legend Editor to create a unique value legend symbolizing each type of road with a different line symbol. In this manner, you can identify and correctly label each road with the appropriate highway symbol.

✓ **TIP:** *You can globally change much of the appearance of the highway symbols by setting the Text and Label defaults under the Graphics menu before you begin to label highways. In the Default Settings, click on the icon for a highway label tool, and set the shield color, type style, type size, and type color.*

The Auto-label Option—Auto-labeling One Theme

If you are dealing with dozens of text labels, the process of labeling becomes increasingly tedious and frustrating. Because the discipline of cartography moved onto computers, one major goal has been to avoid having to handle each piece of type individually, and therefore greatly speed up the labeling process. The Auto-label option in Arc-View is one such attempt, and actually works fairly well. Auto-label does not perfectly place each piece of type, but when used correctly can greatly speed up labeling. The Auto-label option is shown in the following illustration.

The Auto-label feature.

The Auto-label option is found under the Theme menu and operates on one TOC active theme at a time. Before running Auto-label, make sure to choose a reference scale based on final print size of your map. This will help determine the point size needed for readability of the names of the map features you are about to auto-label. Choose a point size and type style before running the Auto-label option by opening the Font palette from the Symbol Window under the Window menu.

✓ **TIP:** *For themes with few features that need labeling, you can set text in a larger type size, without crowding the map. For themes with many feature labels, set text in a smaller point size to ensure that type will not overwhelm the map. To label subsets of features within one theme, copy and paste the theme in the TOC and use the Query tool in the Theme Properties dialog to select a subset.*

The Auto-label dialog box has a variety of useful options. First, choose the Label field from the Theme's attribute table to base the text label on. Then you are presented with a choice of two different means of labeling the theme. "Use Theme's Text Label Placement Property" defaults to the Text Label preferences set in the Theme Properties dialog box. If this radio button is checked, no other refinements are offered. An example of auto-labeling is shown in the following illustration.

Example of auto-labeling.

The "Find Best Label Placement" feature uses different positioning guidelines for text labels, depending on the whether the active theme is a point, line, or polygon. With point themes, text placed with "Find Best Label Placement" will not overlap another point symbol or another piece of text, unless the Allow Overlapping Labels option is checked. For polygon themes, ArcView tries to place the text near the center of the feature, but the type will often extend beyond the polygon's boundaries.

When used with line themes, the best placement feature will prevent a label from crossing its own line or another label, but the label may cross other lines. Line themes activate yet another option box,

Remove Duplicates, which has to do with the way GIS keep line attributes in their databases. Lines are rarely continuous features with only one record per line. Instead, lines such as roads or rivers are broken in dozens if not hundreds of segments, with each segment having its own record in the line attribute table, and as a result, its own possible label.

Roads in an urban area are typically broken into segments that extend between each intersection. Rivers may be broken into segments identifying reaches between each tributary or side stream. As a result, if you auto-label an urban road theme without Remove Duplicates checked, you can end up with hundreds of the same name identifying just one road.

Line themes also have the Line Label Position option of "above," "below," or "on" the line feature. For river or road names, use "above" or "below" for positioning. For placing highway symbols, use the "on" option.

✗ **WARNING:** *If you do not enable Allow Overlapping Labels, Auto-label may choose not to label several features in a theme. It is almost always best to enable Allow Overlapping Labels, and then physically move the overlapping labels with the Pointer tool afterward. ArcView makes it easy to identify overlapping labels: they always appear in a bizarre color, frequently a lime green. After you have moved the label and still have it selected, use the Text Color option in the Color palette to change the lime green back to a more typical black or gray text color.*

At the bottom of the Auto-label dialog are two more check boxes. The Scale Labels box ensures that text stays the same point size as you zoom in and out in the view, and by default is checked "on." The handy Label Only Features in the View Extent box allows you to zoom to a portion of the view, and auto-label only the features visible in the window and not the entire theme. The default for this option is "off."

The Auto-label feature takes some getting used to but is worth experimenting with. Try varying type sizes in the Font palette before using the auto-label feature and then playing with the Use Theme's Text Label Placement Property versus Find Best Label Placement options. Once you get a feel for how this tool works, you will find yourself using it more often, and realizing significant time savings when it comes to labeling your map.

The Multi-theme Auto-label Extension

For the brave user who has mastered the Auto-label feature in basic ArcView, hidden in the ArcView samples directory is the Multi-Theme Auto-Labeler extension, *multhlab.avx*. This freeware extension allows you to auto-label several themes at once by setting the importance level of the theme's features and labels.

With ArcView off, copy the actual extension file, *multhlab.avx*, from the samples directory to the Ext32 directory, which is found normally at *c:\esri\av_gis30\arcview\ext32*. Start ArcView and load the Multi-Theme Auto-Labeler in the Extensions dialog box under the File menu. The Multi-Theme Auto-Labeler appears as an option at the bottom of the Theme menu in the view.

Next, determine which themes you want to label, and make all of them active at once by single clicking on their names in the TOC with the <Shift> key held down. Make sure to be zoomed in or out to what you want to use as your reference scale for labeling; that is, the scale you will print your final map from in the layout. Now you are ready to launch the Multi-Theme Auto-Labeler, shown in the following illustration.

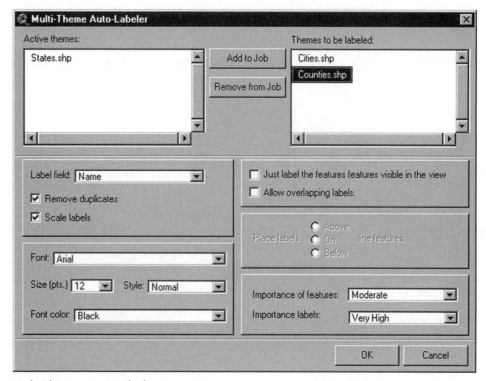

Multi-Theme Auto-Labeler extension.

The Multi-Theme Auto-Labeler dialog box has many of the same options of its simpler predecessor, the Auto-Label feature, with a few new and more powerful tools added. At the top of the dialog box is a window list of the active themes in the view, and you are asked which themes to Add to Job, which moves them to the "Themes to be labeled" window. Below these two windows are a series of options, some of which can be applied to all themes that are to be labeled at once, and others that need to be applied individually.

Select a theme in the "Themes to be labeled" window, and set its Label field on the line to the lower left. If it is a line theme, make sure to check the "Remove duplicates" box. Always check the "Scale labels" box.

Unlike the basic Auto-Label feature, the Multi-Labeler has a handy Font Palette box built in, in the lower left corner. This saves time when setting the type style and the type size, eliminating the need to open the Font palette from the Symbol Window. Identical with the basic Auto-Label feature are the "Just label features in the View," "Allow Overlapping Labels," and "Place labels Above, On or Below for Line" options.

The real power of the Multi-Labeler extension lies in the Importance option found in the lower right corner of the dialog box. Here you can set the importance of features, and the importance of labels at four levels: low, moderate, high, and very high. The importance of features and the importance of labels must be set individually for each of the themes to be labeled. The importance rating for features determines how labels will be placed as to not obscure features on the map, such as the names of towns not overlapping road or river features. An importance rating of Low indicates that ArcView will allow text to overlap map features of active themes added to the Multi-Labeler, while a rating of Very High will prevent text-feature overlap.

The importance rating for labels works in the same manner, but for pieces of text only. An importance rating of Low for labels will result in many overlapping pieces of text, whereas a rating of Very High will result in no overlapping text. The outcome of a Very High importance rating may be that several features are not labeled at all; therefore, it is a good idea to experiment with a High or Moderate setting.

Remember, the point of using tools such as the Multi-Labeler is not to have every piece of type placed perfectly, but to cut down on the time it takes to place large amounts of type. You will always have to move some pieces of text to make your map labeling readable and understandable.

✓ **TIP:** *If you are finishing the map in another graphics program such as Adobe Illustrator or MacroMedia's Freehand, and need to have editable text, export the file as "Placeable WMF" from the view's File menu. This preserves the text objects as editable text when the Windows metafile is opened in a PostScript drawing program.*

Labeling Margin Information

Naming the features in the map efficiently and creating a useful map legend constitute the real heavy lifting of the map labeling process. However, additional text, referred to as "margin information," adds real value to your map product. Some margin information can be entered in the view, whereas other text is appropriately added later in the ArcView layout document.

An example of such margin information is the map title. As a default, ArcView uses for a map title the title of the view window the ArcView layout is based on. Few people will want their final map titled View1or View2! To change this, in the view, go to Properties under the View menu and type in a map title on the Name line.

Do this early in the map production process, and you will find the choice of a descriptive title to be helpful in focusing the other labeling efforts within the view. After all, a map title such as 1999 Annual Sales Figures in California conveys much more information than View1 or just California Sales. Titles can be replaced or altered by using the Text tool in the layout.

Frequently, you will want to add other blocks of text, including a source statement, to your map. Larger amounts of text can be added as a graphic block in the layout using the Text tool, but typing multiple paragraphs directly into the dialog box with this tool can be very clumsy.

✓ **TIP:** *Formatted text can be copied from a word processing program into the Text tool dialog box in ArcView. By using a word processor, you can precisely format entire paragraphs of marginal text for your map, and copy and paste them using keyboard commands (<Control+C> and <Control+V>) into the ArcView Text tool dialog box. This also allows you to save important information, such as your company's address, in the text file in the word processing document for use in other maps.*

Beyond the Basics—Advanced Labeling

When the basic tools in ArcView fail to meet your map's labeling needs, you are faced with a choice: purchase a specialized labeling extension or export the map to a dedicated graphics program. Either choice will cost you money, and is dependent on how many labels you need to place on the map, as well as on what your final map product will be.

If you need to place several hundred labels or just want to speed up labeling tasks, check out an ArcView extension called AVALabel 2.3, the advanced labeling tools of which are shown in the following illustration. If you need to precisely place thousands of text labels for a high-end map product that will be published in an atlas or book, export the view or layout document to a graphics program such as Adobe Illustrator or MacroMedia Freehand, and use an extension to those programs called MaPublisher.

AVALabel, which stands for ArcView Advanced Label, is an outstanding extension to ArcView from MapLogic Corporation, and can be found on the web at *http://www.maplogic.com*. AVALabel fixes many of the shortcomings of the basic ArcView label tools, and provides several time-saving shortcuts for the labeling process. The following material describes these features.

Advanced Labels feature in AVALabel 2.3.

AVALabel solves the vexing label reference scale problem by allowing you to specify a reference scale for the themes in the view, and labeling features at that same scale, whether you have zoomed in or out of the window. This feature alone is worth the price of the extension! In basic ArcView, you are forced to remember the reference scale or physically write it down on a piece of paper, and return to that zoom level before placing additional text.

AVALabel easily allows the combination of multiple fields from an attribute table to be used in a single piece of text. This can be done in basic ArcView only, using the Avenue commands in the Table's Field Calculator to create a new field. The Fixed Scale Label tools of AVALabel 2.3 are shown in the following illustration.

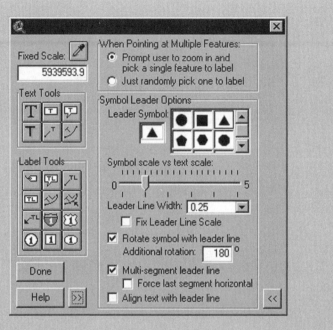

Fixed Scale label tools in AVALabel 2.3.

After you have placed a label, AVALabel provides you with a wealth of useful features for manipulating the label. The Nudge tools allow you to move pieces of text by specified increments, such as points or inches, giving a degree of positioning control that is impossible using just ArcView's Pointer tool to drag text around the view.

Like a word processor, AVALabel offers a search and replace function, permitting view-wide searches and editing of feature names. Once you are satisfied with a set of labels, AVALabel will save that set to a file, and allow you to use it again and again in different projects and views.

AVALabel is by far one of the most useful ArcView extensions available, and should be part of every cartographer's digital toolbox. However, AVALabel does not solve the problem of "ArcView project bloat," which creates huge, multi-megabyte project files caused by hundreds or thousands of graphic text objects in the ArcView project. When you reach this stage, or when you need to prepare the map for high-end publishing, it is time to consider moving the map to a dedicated graphics program.

Dedicated graphics programs such as Adobe Illustrator or Macromedia Freehand provide much finer control over type and color than is found in ArcView, and generate bombproof PostScript files, which are the standard file structure for book and magazine publishing.

Moving an ArcView map, exported as an Encapsulated PostScript file (EPS), to a graphics program used to be a cumbersome task. Today, a software program called MaPublisher 3.5 bridges the gap between GIS and CAD programs and graphics programs by functioning as a "mini-Arc-View" within the graphics program. MaPublisher 3.5, produced by Avenza, is an extension to Illustrator and Freehand, and imports and manipulates ArcView shapefiles in these two programs. This means that you can label map features within the graphics program from the shapefile's attribute table, and take advantage of the refined typographic control found there. MaPublisher is shown in the following illustration.

Labeling map features in MaPublisher 3.5.

Of course, the disadvantage of this approach is that you may need to buy and learn yet another software program, in addition to ArcView, which can be a daunting task. However, depending on the amount of text and type of map product you need, Illustrator or Freehand plus MaPublisher may be the most efficient approach to take. For more information on MaPublisher, visit their web site at *http://www.avenza.com.*

Summary

Labeling map features (placing type on a map) is a complex and time-consuming task. ArcView, including a couple of extensions, provides a variety of tools to speed up the task, but by no means makes the labeling process a trivial exercise. Make sure to consider the amount of labeling your map needs, and then budget the time in the map-making process to do it right.

Chapter 7

Map Projections and Map Scale in the View Document

Miles 500

One of the most confusing aspects of mapmaking is the use of projections. Traditionally, map projections required working through a lot of elaborate equations to get the proportions of curved land masses correct when they were squashed flat on a piece of paper. It would be a lot easier for cartographers if we did live on a flat Earth!

Computers have helped tremendously with the math, but have at times added problems. How do you know what map projection to use, out of the hundreds available? And how do you know if the software is doing the projection correctly? ArcView, for example, has two built-in tools for changing projections, one of which works much better than the other, whereas a third add-on extension beats both of them for ease of use. This chapter is intended to clear up much of the confusion, and provide guidance on how to project map data in ArcView and what projection to use for your needs.

Why Do We Need Map Projections?

Map projections are the cartographer's tool for taking a curved surface of reality, reducing it in scale, and placing it on a flat surface, such as a piece of paper or a flat computer screen. However, no tool is perfect, and every map projection will distort one or more of the following attributes on a map: shape, area, distance, or direction. Only on the curved surface of a globe will all four attributes appear undistorted, but globes are difficult to produce and clumsy to carry when traveling. Flat maps are a must.

Projections are designed to accurately depict one or two of those attributes, frequently at the expense of the others. It is the cartographer's job to choose projection tools carefully, depending on what the map is needed for and the geographic extent of the map. The classic case in point is the use of the Mercator projection for depicting the entire Earth.

Gerardus Mercator lived in sixteenth-century Europe, and devised a world map for use by mariners sailing between Europe and the Americas. The Mercator map treated Earth as a cylinder, whose surface could be rolled flat on a chart table on the deck of a ship. Any straight line between two points on the Mercator map was a line of constant direction called a rhumb line. Mariners loved the convenience of the Mercator map for navigation, especially in the mid to lower latitudes, where most of the east-west commerce took place between continents.

The Mercator map was so popular that it became a fixture in schools when teaching geography, long after the sixteenth century had passed. Many school children educated in the twentieth century absorbed their main view of the world from the familiar Mercator map hung on the classroom wall. Unfortunately, a Mercator map is not designed to give people a realistic view of the world, unless of course they yearn to grow up to be sixteenth-century sea captains.

The Mercator projection significantly exaggerates a polygon feature's area the further north or south of the equator you move. Because much of the continental landmasses are in the mid to high latitudes of the Northern Hemisphere, North America and Asia loom larger than life. The frozen island of Greenland appears as big as South America, whereas in reality South America is eight times larger.

In the early 1980s, one black scholar pointed out that the Mercator projection minimized the size, and therefore the importance, of the continent of Africa (which straddles the equator) compared to North America and Europe. Africa appears smaller than North America on a Mercator map, whereas in reality it is 20 percent larger. Cognizant of this controversy, the National Geographic Society adopted the Robinson projection as its standard for world maps, which gives a much better view of the comparative sizes of the Earth's continents.

Perceptions and misperceptions created by maps can indeed influence the way people think about the world. The following chart lists common map projections and gives an idea of their uses and scales of use.

Common Map Projections and Their Uses

Adapted from the poster, "Map Projections," U.S. Department of the Interior, U.S. Geological Survey

● = Yes   SS = Small Scale
○ = Partly   LS = Large Scale

Projection	Type	Conformal	Equal area	Equidistant	True direction	Perspective	Compromise	Straight rhumbs	World	Hemisphere	Continent/Ocean	Region/Sea	Medium Scale	Large Scale	Topographic Maps	Geological Maps	Thematic Maps	Presentations	Navigation	USGS Maps
Globe	Sphere	●	●	●	●				●									●	●	
Mercator	Cylindrical	●			○			●	○		●				●	●			●	●
Transverse Mercator	Cylindrical	●									●	●	●	●	●	●				●
Oblique Mercator	Cylindrical	●										●	●	●	●	●				●
Space Oblique Mercator	Cylindrical	●												●						●
Miller Cylindrical	Cylindrical						●		●								●			
Robinson	Pseudocylindrical						●		●								●	●		
Sinusoidal Equal Area	Pseudocylindrical	●	○						●	●							●			●
Orthographic	Azimuthal				○	●				○										●
Stereographic	Azimuthal	●			○	●				●	●	●	●	●	●	●				●
Gnomonic	Azimuthal				○	●						○						●	●	
Azimuthal Equidistant	Azimuthal			○	○				○	●	●	●		○	●					●
Lambert Azimuthal Equal Area	Azimuthal		●	○						●	●	●					●	●		●
Albers Equal Area Conic	Conic		●								●	●	●				●	●		●
Lambert Conformal Conic	Conic	●			○						●	●	●	●	●	●				●
Equidistant Conic (Simple Conic)	Conic			○							●	●								
Polyconic	Conic				○		●						○	○						●
Bipolar Oblique Conic Conformal	Conic	●									●						●			●

Projection Terminology and Concepts

Before introducing ArcView's projection tools, you need a grounding in the arcane terminology of map projections, in order to minimize confusion later. Once the terminology and concepts are mastered, or at least partially understood, it becomes much easier to select a projection to meet your mapping needs.

Projection Types

There are three main projection types based on the process of expanding the globe onto a flat surface: cylindrical, azimuthal (planar), and conic. In addition, there are several altered projection types that tweak the main types. These include pseudocylindrical projections and modified projections. A tangent is the single line where these geometric objects intersect the curved surface of the sphere.

Secants are the two lines where these geometric objects intersect the curved surface of the sphere more than once. In general, distortion of map features increases as the distance from these lines of intersection increases. The following illustration shows tangents and secants.

Tangents and secants.

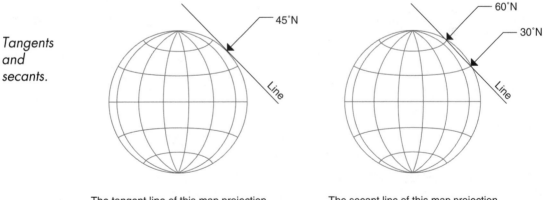

The tangent line of this map projection intersects the globe at one point.

The secant line of this map projection intersects the globe at two points.

With a cylindrical projection, a hypothetical cylinder is slid over the globe, and the globe is projected onto its surface. The cylinder need not be vertical north to south, but can be horizontal or oblique. The Mercator projection is the best-known example of a cylindrical projection. The following illustration shows cylinders, cones, and planar surfaces.

Cylinders, cones, and planar surfaces.

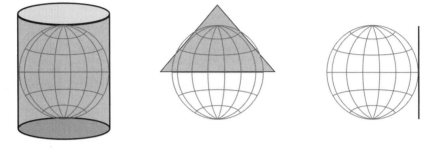

A normal cylindrical projection.   A secant conic projection.   A planar equatorial projection.

In a conic projection, a cone is slipped over the globe. Conic projections are particularly good at minimizing distortion for areas in the mid latitudes of the northern or southern hemispheres. The tangent or secant lines where the cone intersects the globe are called standard parallels of latitude. Albers Equal Area and Lambert Conformal are common conic projections. An Albers Equal Area conic projection is shown in the facing illustration (top).

An azimuthal projection type utilizes a flat (planar) surface intersecting the globe, and the globe is mapped onto the flat surface. An orthographic projection and a Lambert Equal Area are examples of azimuthal projections. A Lambert Equal Area azimuthal projection is shown in the facing illustration (bottom).

Albers Equal
Area conic
projection.

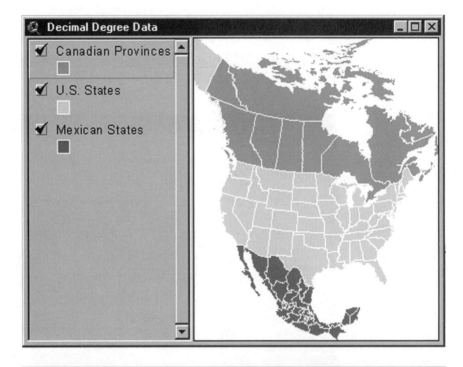

A Lambert Equal
Area azimuthal
(planar)
projection.

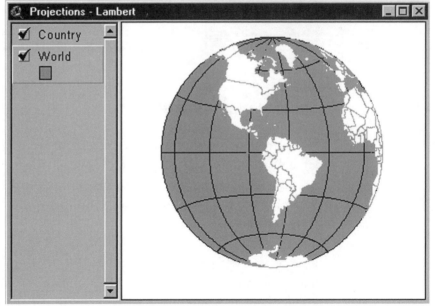

Altered projection types include the Robinson projection, which is a compromise between conic and planar projections called a pseudocylindrical projection. A Robinson projection is shown in the following illustration.

A Robinson projection.

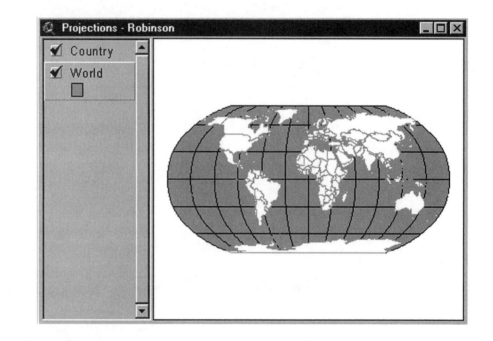

Spheres and Spheroids

Not only would mapping be easier if Earth were flat, it would be easier if Earth were a perfect sphere. Instead of a perfect sphere, Earth resembles a squashed beach ball or even a rotund pear. This is because Earth rotates rapidly, causing the globe to expand at the equator and flatten at the poles.

Due to the vagaries of continental drift, more of the planet's landmasses are in the Northern Hemisphere, whereas the Southern Hemisphere contains more ocean. Continental rock is denser than water, and the huge landmasses of Eurasia and North America squish the Northern Hemisphere inward. Strangely enough, as a result, the South Pole is closer to the equator than the North Pole.

For small-scale maps of 1:5,000,000 scale, the Earth can be treated as a sphere. For larger scale maps of 1:1,000,000 and greater, the squashed beach ball effect has to be taken into account. In order to take these factors into consideration, cartographers use ellipsoids called spheroids to model the surface of the globe. Spheres and spheroids are shown in the following illustration.

Spheres and spheroids.

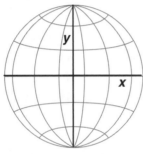

The globe as a perfect sphere.   The globe as a spheroid.

In North America, two spheroids have been in common usage: Clarke's 1886 spheroid and GRS80. The properties of Clarke's 1886 spheroid were surveyed and calculated manually over several decades, whereas the newer version of GRS80 (Geodetic Reference System 1980) takes into account satellite observations and computer calculations. Spheroids are used as the references for geodetic datums, such as NAD27, NAD83, and WGS84.

Datums

Geodetic datums use a reference spheroid to describe the shape of Earth, and provide a frame of reference for designing coordinate systems for locating features on Earth. There are local datums, used for entire continents or portions of continents, and there are global datums.

Two local datums in use for North America are the North American Datum of 1927 (NAD27) and the North American Datum of 1983 (NAD83). Many older maps were based on the NAD27 frame of reference. A lot of GIS data was digitized from older maps, and as a result, a lot of digital data retains the NAD27 datum. The newer NAD83 datum is based on the Geodetic Reference System 1980 (GRS80) and is the primary datum in use for the United States today. The World Geodetic System of 1984 (WGS84) is the datum used by the global positioning system of satellites used by GPS receivers to locate features.

Changing a datum requires moving from one entire reference system to another. This entails performing a geographic transformation or datum conversion. For North America, the North American Datum Conversion (NADCON) is used to move data from the old NAD27 to the newer NAD83 datum. This is discussed further in the section on using the ArcView Projection utility later in this chapter.

Coordinate Systems

Coordinate systems use x, y, and sometimes z values to locate a point on the earth's surface. Coordinate systems are based on datums, and can either be geographic or projected. Geographic coordinate systems use the curved surface of the globe as their base, whereas projected coordinate systems involve the flat surface of a map.

Geographic coordinate systems use latitude and longitude to find locations on the globe, which are usually recorded as degrees, minutes, and seconds. The origin or zero point for latitude is the equator, whereas the origin point for longitude is the prime meridian in Greenwich, England. Lines of latitude are often called parallels, because they are parallel with one another between the poles and the equator. Lines of longitude are referred to as meridians, and converge toward the poles. As a result, degrees of latitude or longitude do not represent the same distance everywhere on the globe. The grid formed by latitude and longitude lines is called a graticule.

In order to bring tabular point data into ArcView in a geographic projection (from a GPS unit, for example), it must be stored as decimal degrees. Latitude is stored as the y coordinate in one field, which is a positive number north of the equator and a negative number south of it. Longitude is stored in another field as the x coordinate, which is a negative number west of the prime meridian (in North America) and a positive number east of it (most of Europe and Asia).

➥ **NOTE:** *For information on how to store tabular point data as decimal degrees, see the scripts* dms2dd.ave *and* dd2dms.ave *in the sample script directory in your ArcView install location on your hard drive.*

Projected coordinate systems create a grid of uniform measures on the flat surface of a map, starting at an arbitrary origin point. Many countries have established their own national grid systems, such as Great Britain, whereas others use smaller grids for their political units, such as the State Plane coordinate system in the United States. A global projected coordinate system is the Universal Transverse Mercator (UTM) zone system. Coordinate systems are described in more detail under the section "Projections in ArcView" later in this chapter.

Projection Parameters

When entering a projection in ArcView, you will be asked to fill in projection parameters. Projection parameters control precisely how the section of the globe you are interested in gets flattened onto the map. Depending on what projection tool you are using, you will have to establish between two and ten parameters. ArcView's projection parameters are shown in the following illustration, and are described in the text that follows.

Projection parameters in the Projection Properties feature.

Projection: Name of the projection you will use to convert the curved surface to a flat surface.

Spheroid: Ellipsoid that will be used to represent the earth's curvature.

Central Meridian: The line of longitude in the center of the projected area, which serves as a point of origin for the *x* coordinate of the projection.

Reference Latitude: The latitude of the projection's origin. The reference latitude and the central meridian help create the origin of the measured grid that will define the map. A false origin is frequently set to west and south of the real origin, in order to keep the map measures in a positive range. With a cylindrical or azimuthal projection, the origin point will be the point of tangency with the cylinder or flat surface.

Standard Parallel 1: For a conic projection, the line on the southern boundary of the map at which the cone intersects the globe (a secant line).

Standard Parallel 2: For a conic projection, the line on the northern boundary of the map at which the cone intersects the globe (another secant line).

False Easting: The measure in feet or meters of the x coordinate offset from the projection's origin.

False Northing: The measure in feet or meters of the y coordinate offset from the projection's origin.

Choosing a Map Projection

After wading through the intricacies of map projections, it might seem impossible to choose the "right" projection. What the "right" projection is depends on the scale of your data, what you need to accomplish with the map, and the shape of your area. The following are several questions to answer.

How large a geographic area do you need to put on the map? For small-scale data covering several continents, try a Robinson, Miller Cylindrical, or an Equal Area cylindrical projection. Do the areas of countries or regions need to be comparative in size? Do you need to measure distances with some degree of accuracy between points? Select an Equidistant projection in this case. In addition, try using small-scale data stored in decimal degrees to allow ease of changing projections with the View Properties projection feature for these situations. (Table 7-1, Common Map Scales and Equivalent Measures, provides help in comparing projection properties to map scale and proper use.)

If you are new to mapping with ArcView and do not have an idea of what projections to use for large-scale mapping, examine published maps of your area. A good world atlas will list the projection and even the parameters used for projections for a map of your country or province. Government maps, such as the 1:24,000 scale series of topographic maps from the U.S. Geological Survey (USGS), can provide clues on projections to adopt for your local area. Many large-scale maps, such as the 7.5-minute series from the USGS, will use a UTM projection for a particular zone to minimize distortion.

In the United States, try using a State Plane projection for data encompassing a single state. State Plane projections take into account the shape of the state in selecting a base projection, with wide areas using a Lambert Conformal projection and tall areas using a Transverse Mercator projection.

Questions of Scale

Map scale is expressed as the ratio between the distance on the map and the related distance on the earth. With large-scale maps, the ratio is closer to one than with small-scale maps. That is, distances on large-scale maps represent fewer real measures on the ground than do distances on small-scale maps. Table 7-1, which follows, presents the

Color Plate 1
Predefined Graduated Color Ramps in ArcView

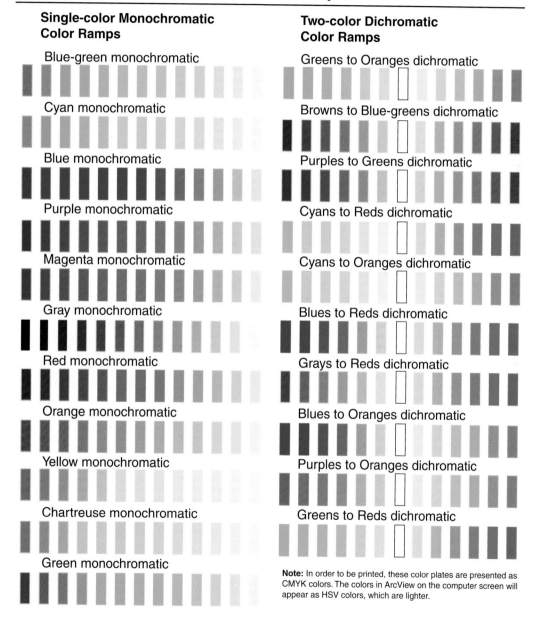

Single-color Monochromatic Color Ramps

Blue-green monochromatic

Cyan monochromatic

Blue monochromatic

Purple monochromatic

Magenta monochromatic

Gray monochromatic

Red monochromatic

Orange monochromatic

Yellow monochromatic

Chartreuse monochromatic

Green monochromatic

Two-color Dichromatic Color Ramps

Greens to Oranges dichromatic

Browns to Blue-greens dichromatic

Purples to Greens dichromatic

Cyans to Reds dichromatic

Cyans to Oranges dichromatic

Blues to Reds dichromatic

Grays to Reds dichromatic

Blues to Oranges dichromatic

Purples to Oranges dichromatic

Greens to Reds dichromatic

Note: In order to be printed, these color plates are presented as CMYK colors. The colors in ArcView on the computer screen will appear as HSV colors, which are lighter.

Color Plate 2
Predefined Graduated Color Ramps and Palettes in AV

Two- and Three-color Ramps

Yellow to Orange to Red

Beige to Brown

Red to Purple to Blue

Green to Cyan to Blue

Yellow to Green to Dark Blue

Specialized Color Ramps

Elevations #1

Elevations #2

Sea Floor Elevation

Full Spectrum

Precipitation

Temperature

Land Cover #1

Land Cover #2

Artist.avp Color Palette

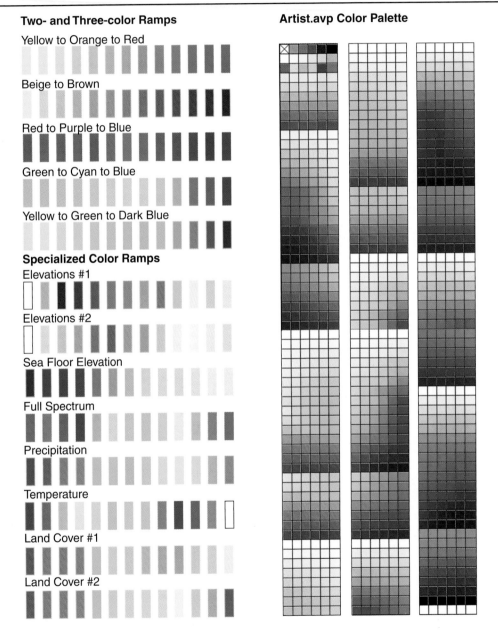

Color Plate 3
Predefined Color Schemes in ArcView

Color Families

Bountiful Harvest

Pastels

Minerals

Fruits and Vegetables

Cool Tones

Warm Tones

Autumn Leaves

Equatorial Rain Forest

The High Seas

Color Plate 4
Process Color Test Chart for ArcView

Note: To print this chart, go to the calibrat.apr file in the ArcView Samples directory on your computer.

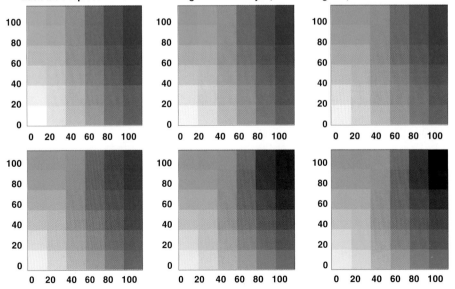

CMYK Color Tables
Each table represents 20% incrementing columns of Cyan, rows of Magenta, and blocks of Yellow.

Color Plate 5
Graduated Color Legend Types in ArcView

Plate 5a - Population Growth in the Lower 48 States by State 1990-1997. The default graduated color legend using a natural breaks classification with five classes shows only general population trends.

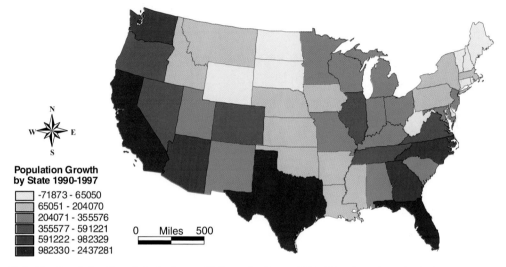

**Population Growth
by State 1990-1997**

- -71873 - 65050
- 65051 - 204070
- 204071 - 355576
- 355577 - 591221
- 591222 - 982329
- 982330 - 2437281

0   Miles   500

Plate 5b - Population Growth in the Lower 48 States by County1990-1997. A graduated color legend with a standard deviation classification quickly pinpoints what parts of the country have high growth. County polygon outlines are light gray to avoid cluttering the map.

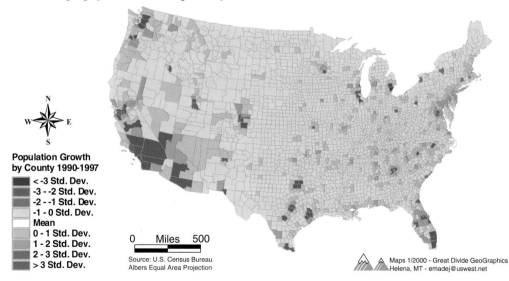

**Population Growth
by County 1990-1997**

- < -3 Std. Dev.
- -3 - -2 Std. Dev.
- -2 - -1 Std. Dev.
- -1 - 0 Std. Dev.
- Mean
- 0 - 1 Std. Dev.
- 1 - 2 Std. Dev.
- 2 - 3 Std. Dev.
- > 3 Std. Dev.

0   Miles   500

Source: U.S. Census Bureau
Albers Equal Area Projection

Maps 1/2000 - Great Divide GeoGraphics
Helena, MT - emadej@uswest.net

Color Plate 6
Chart Legend Type in ArcView

Chart Legend - Average annual wind speed for selected cities is shown in this bivariate chart legend. The size of the pie chart is scaled to the average annual wind speed for each city, while the slices reflect the average wind speed by season. The legend was created using the graphic tools in the ArcView Layout. Notice that the cities east of the Rocky Mountains and those in the Northwest are much windier than the cities of the Southwest.

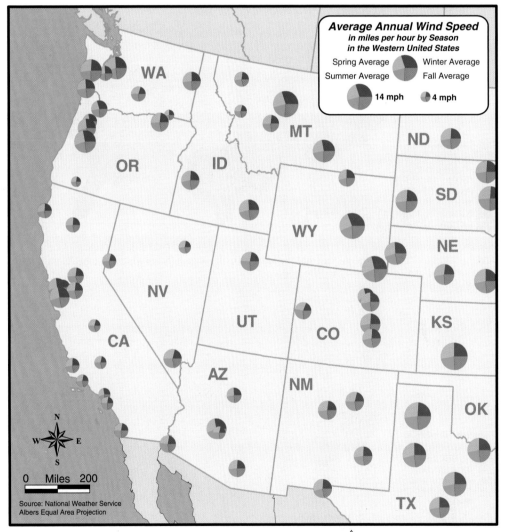

Color Plate 7
Using Buffers for Cartographic Effects in ArcView

Water Buffers - Use the Create Buffers feature in the Theme menu to make great cartographic effects. Here the land areas of Washington State and British Columbia are buffered as outside rings 2.5 km apart, and filled with light to dark blue, to give the impression of deepening water.

Strait of Georgia

Vancouver
Delta

BRITISH COLUMBIA
CANADA
UNITED STATES
WASHINGTON

VANCOUVER ISLAND

Bellingham

Sidney

Victoria

Friday Harbor

Anacortes

Strait of Juan de Fuca

Port Angeles

Ferry Routes
in Southeastern British Columbia
and Northern Washington State

0   Miles   40

N
W   E
S

Everett

Source: National Geographic Road Atlas
Washington State Plane North Projection

Map 1/2000 - Great Divide GeoGraphics
Helena, MT - emadej@uswest.net

Color Plate 8
Using Buffers for Cartographic Effects in ArcView

Land Buffers - Use the Create Buffers feature in the Theme menu to make political jurisdictions stand out, while minimizing linework and clutter on the map. The county polygons of Washington, Idaho, and Montana are buffered as one inside ring and filled with a darker color than the polygon fills. Loading the *artist.avp* ArcView color palette helps greatly in choosing the correct colors.

The Inland Empire
Spokane, Washington,
and Northern Idaho

MONTANA
IDAHO

395

95

2

Spokane

KOOTENAI

Coeur d'Alene

90

Opportunity

SPOKANE

90

95

195

WASHINGTON
IDAHO

BENEWAH

WHITMAN

LATAH

Pullman

Moscow

0   Miles   20

N
W   E
S

Source: ESRI Data. Washington
State Plane North Projection

Map 1/2000 - Great Divide GeoGraphics
Helena, MT - emadej@uswest.net

Color Plate 9
Graduated Symbol Types in ArcView

Graduated Point Symbol Legend - Graduated symbol maps are tricky to do properly in ArcView, and maps 9a and 9b show some crucial differences. Both maps use a shaded relief TIFF image as a base, assembled in ArcInfo by the U.S. Geological Survey, providing a great perspective for the terrain in the Yellowstone area.

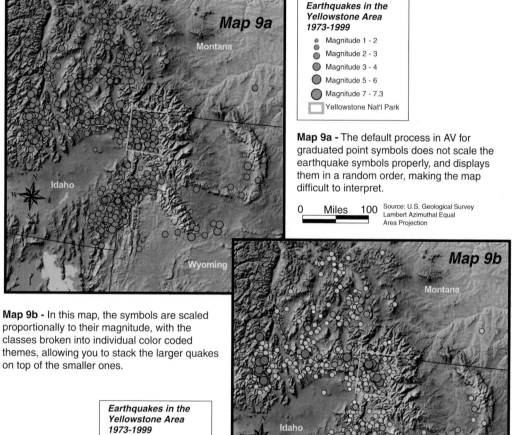

Earthquakes in the Yellowstone Area 1973-1999
- Magnitude 1 - 2
- Magnitude 2 - 3
- Magnitude 3 - 4
- Magnitude 5 - 6
- Magnitude 7 - 7.3
- Yellowstone Nat'l Park

Map 9a - The default process in AV for graduated point symbols does not scale the earthquake symbols properly, and displays them in a random order, making the map difficult to interpret.

0 Miles 100

Source: U.S. Geological Survey
Lambert Azimuthal Equal
Area Projection

Map 9b - In this map, the symbols are scaled proportionally to their magnitude, with the classes broken into individual color coded themes, allowing you to stack the larger quakes on top of the smaller ones.

Earthquakes in the Yellowstone Area 1973-1999
- Magnitude 5 - 7.3
- Magnitude 5 - 6
- Magnitude 4 - 5
- Magnitude 3 - 4
- Magnitude 2 - 3
- Magnitude 1 - 2
- Yellowstone Nat'l Park

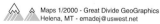

Maps 1/2000 - Great Divide GeoGraphics
Helena, MT - emadej@uswest.net

Color Plate 10
Advanced Graduated Color Legends in ArcView

Western Annual Precipitation - This map uses contours (polygons) to show ranges of precipitation amounts. Because the American West is primarily dry with a few very wet areas, most default color ramps were insufficient. A custom ramp extending from light tan to dark blue, and forced to pass through green, was constructed using the *artist.avp* palette in the Legend Editor. A special extension, *Legend-Smooth-Continuous.avx*, was used to construct the map legend.

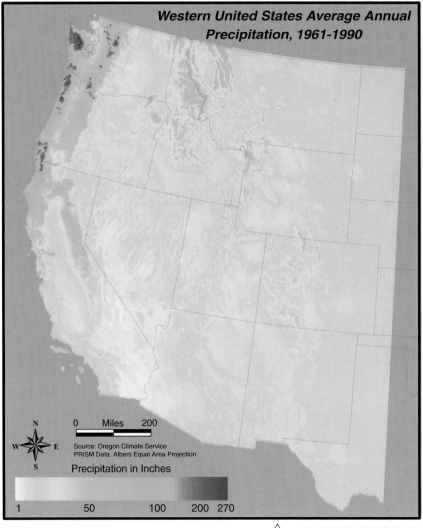

Color Plate 11
Using Digital Raster Graphics in ArcView

Topo Maps as Backgrounds - Digital Raster Graphics or DRGs are scanned, georegistered images of USGS 1:250K, 1:100K, and 1:24K topographic maps that add a wealth of information as backgrounds for ArcView maps. Care must be taken with overlying themes as to not obscure too much of the background DRG, and AV shapefiles must frequently be reprojected to match the DRG's original projection. Here parks and city boundaries are outlined on the Helena, Montana, 7.5' quadrangle.

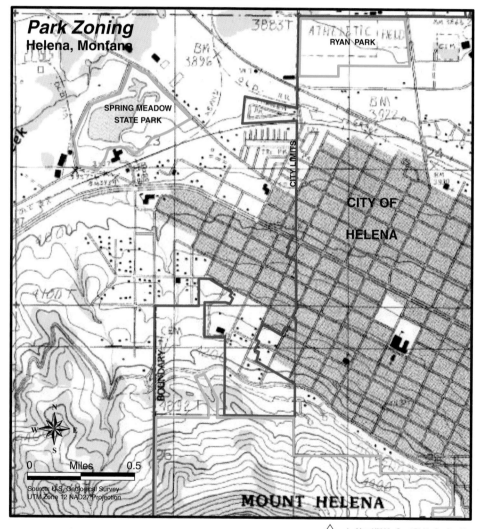

Color Plate 12
Using Digital Ortho Quads in ArcView

Aerial Photos as Backgrounds - Digital Ortho Quarter Quads or DOQQs are high-quality aerial photos that are georegistered to match a quarter of a particular USGS 7.5' topopgraphic map. The DOQQs provide a tremendous amount of background detail for ArcView maps, and present many of the same problems as topo map DRGs with printing and projections. In this map, park boundaries are shown in a light green transparent pattern, using the *carto.avp* palette, and possible flood zones are outlined with a 200-meter buffer (shown in blue) using the Create Buffers feature. Compare map detail with plate 11.

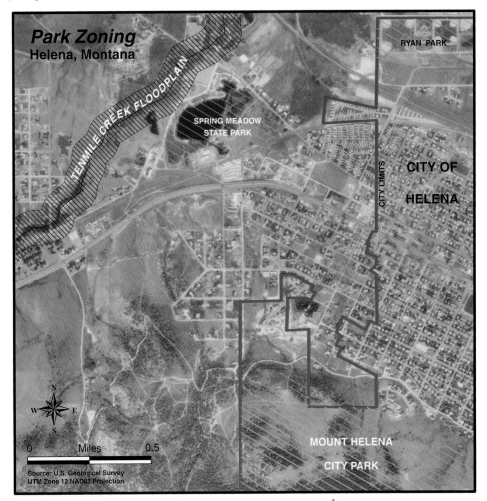

Color Plate 13
Shaded Relief Maps in ArcView Spatial Analyst

Shaded Relief Maps - Producing relief maps in AV Spatial Analyst requires a few tricks. You can start with a USGS DEM imported into ArcView as a GRID. Apply elevation ramps or custom palettes in the Legend Editor. Construct a hillshade from Spatial Analyst's Surface menu. Go back to the DEM's Legend Editor and use the Advanced button to select the new hillshade as the DEM's brightness theme and apply the legend.

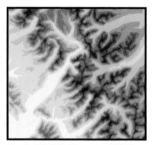

Basic USGS 30-meter DEM, saved as a GRID.

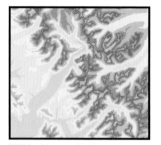

DEM with standard Elevation #2 color ramp from ArcView's Legend Editor.

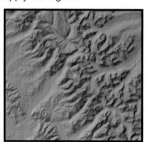

DEM with Elevation #2 color ramp and default hillshade, of azimuth 315°, 45° sun altitude.

DEM with custom hillshade, azimuth 45°, 65° sun angle, 23 elevation classes, and custom *ddvelev8b.avp* palette "California Dry - Sunny" legend from Jim Mossman.

The relief map at right was created by using a custom elevation color palette created for an area with a greater elevation range (California) and adapting it to the 3,000' to 10,000' range of Glacier National Park. An unusual sun position and angle (from the northeast) was used to create the hillshade, and a continuous legend was constructed using the extension *Legend-Smooth-Continuous.avx*. 1:100,000 scale hydrologic features were drawn on top of the 1:24,000 scale elevation GRID.

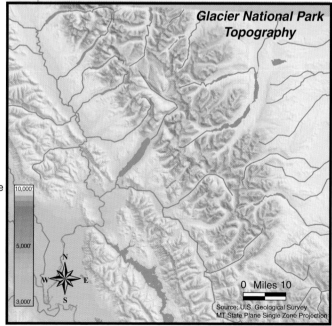

Glacier National Park Topography

10,000'

5,000'

3,000'

0 Miles 10

Source: U.S. Geological Survey
MT State Plane Single Zone Projection

Maps 1/2000 - Great Divide GeoGraphics
Helena, MT - emadej@uswest.net

Color Plate 14
Advanced Hillshades in ArcView Spatial Analyst

Advanced Hillshades - Hillshades can be used to show depth or elevation for a variety of surface features. Here, the custom hillshade of azimuth 45˚, 65˚ sun angle derived from a DEM is used with a landcover theme showing current vegetation communities for the Glacier National Park bioregion.

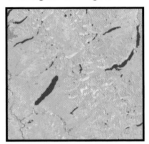

90-meter pixel vegetation theme saved as a GRID.

Custom hillshade derived from 30-meter digital elevation model.

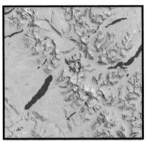

Vegetation theme with hillshade applied as the brightness theme in the Advanced feature.

Vegetation for Glacier National Park
and surrounding areas

Vegetation types vary widely in the greater Glacier National Park bioregion. The legend is too long to be shown here, but the hillshade helps identify green coniferous forests in the mountainous portion, whereas red urban areas appear in the rolling Flathead Valley in the southwest corner of the map.

The striking close-up map below shows glacial cirques and mountain ridges, with yellow and light blue colors identifying alpine grasslands.

0 Miles 10

N W E S

Source: Montana Gap Analysis Project
Albers Equal Area Projection

Maps 1/2000 - Great Divide GeoGraphics
Helena, MT - emadej@uswest.net

Color Plate 15
Perspective Views with ArcView 3D Analyst

3D Terrain Views - Oblique views of topography are not maps as we know them, but a powerful form of data visualization. A DEM is frequently the base of the 3D model, and other features (rivers, roads, and buildings) can be added. Other themes, such as landcover or aerial photos, can be draped over the DEM.

Glacier DEM with standard Elevation #2 color ramp, zero vertical exaggeration.

Glacier DEM with custom elevation color ramp and hillshade 2x vertical exaggeration.

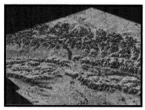

Vegetation theme with hillshade applied as the brightness theme draped over Glacier DEM.

Left View - Close-up of the Lake MacDonald and Logan Pass area of Glacier National Park with Going-to-the-Sun Road, looking to the northeast. Elevations range from 3,500' at the lake, to over 10,000' at the mountain tops. Custom elevation ramp, *ddvelev8b.avp* palette "California Dry - Sunny" legend courtesy of Jim Mossman. All views saved as RGB JPEG images from AV and converted to CMYK TIFF images in Adobe Photoshop.

Right View - Same close-up with vegetation draped over the DEM surface. Both Vegetation and DEM themes are turned on (made visible) in ArcView, and the vegetation theme's 3D properties are set to 60% transparent with shade features checked.

Source: DEM - U.S. Geological Survey
Vegetation - MT Gap Analysis Project
Albers Equal Area Projection

Maps 1/2000 - Great Divide GeoGraphics
Helena, MT - emadej@uswest.net

Color Plate 16
Interactive Maps with ArcView Internet Map Server

AVIMS Maps - Interactive dynamic maps have different visual design considerations than static paper maps. Although created with the same basic software, ArcView IMS maps need to have simpler color schemes using web-safe colors that will show up on most computer monitors.

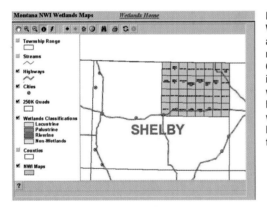

Interactive maps with ArcView can be clean and simple in order to address a particular need, such as this watershed finder. Colored polygons define the major watersheds of Montana. Users can use the Identify tool to query individual watersheds (shown below) or use the Hotlink tool to view a web page with linked data on a specific group working on local watershed issues.

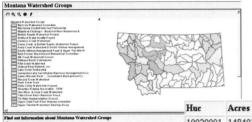

Huc	Acres	Name	WSGroup	Links
10020001	1484038	RED ROCK	Snowline Grazing Association - CRM	SGACRM.htm

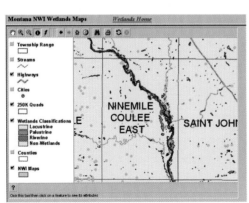

In this more complex example, data from the National Wetlands Inventory is made available for online querying and mapping. The map to the left shows a grid of 7.5' minute maps with wetlands data available for Shelby County. On the lower left, the map user zooms in to an individual 1:24K topo quad, and scale-dependent wetlands data becomes visible for Ninemile Coulee East. Below, a close-up view depicts three types of mapped wetlands in darker colors, and non-wetland areas as light green. The online map user can turn on other themes, such as cities, highways, and streams.

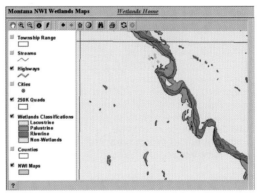

Source: Data from National Wetlands Inventory, U.S. Fish and Wildlife Service
Interactive Maps courtesy Velda Welch, MT Natural Resource Information System
Scale varies from window to window, Albers Equal Area Projection

relationship between map scale and the corresponding measures on the ground.

Table 7-1: Common Map Scales and Equivalent Measures*

Map Scale	1 Inch on Map Equals	1 Mile on Earth Equals	1 cm on Map Equals	1 km on Earth Equals
1:1,000	83.42 feet	63.36 inches	10 meters	100 cm
1:1,200	100	52.80	12	83.33
1:2,000	186.24	31.68	20	50
1:2,400	200	26.40	24	41.66
1:2,500	208.56	25.30	25	40
1:4,800	400	13.20	48	20.83
1:5,000	417.12	12.67	50	20
1:6,000	500	10.56	60	16.66
1:9,000	750	7.04	90	11.11
1:9,600	800	6.60	96	10.42
1:10,000	834.24	6.34	100	10
1:12,000	1000	5.28	120	8.33
1:15,840	1320	4	150	6.32
1:20,000	1668.48	3.17	200	5
1:24,000	2000	2.64	240	4.17
1:25,000	2085.60	2.53	250	4
1:31,680	0.500 miles	2	317	3.16
1:50,000	0.789	1.27	500	2
1:62,500	0.986	1.014	625	1.60
1:63,360	1	1	634	1.58
1:80,000	1.26	0.792	800	1.25
1:100,000	1.58	0.634	1000	1
1:125,000	1.97	0.507	1250	0.80
1:126,720	2	0.5	1267	0.79
1:250,000	3.95	0.253	2500	0.40
1:500,000	7.89	0.127	5000	0.20

Map Scale	1 Inch on Map Equals	1 Mile on Earth Equals	1 cm on Map Equals	1 km on Earth Equals
1:1,000,000	15.78	0.063	10 km	0.10
1:2.000,000	31.56	0.032	20	0.05
1:3,000,000	47.35	0.021	20	0.03
1:5,000,000	78.91	0.013	50	0.02
1:10,000,000	157.82	0.006	100	0.01
1:25,000,000	394.57	0.002	250	0.004

* Adapted from ArcView Help.

Scale Related to Projection

The projection you choose for a map depends in part on the map scale. A Robinson projection, designed for working on a global level, will not work with large-scale data. Likewise, a UTM Zone projection designed for small geographic areas does not work well with areas approaching the average-size state. The following illustration shows a map employing a Washington State Plane North projection.

A small-scale map of Washington State using a Washington State Plane North projection.

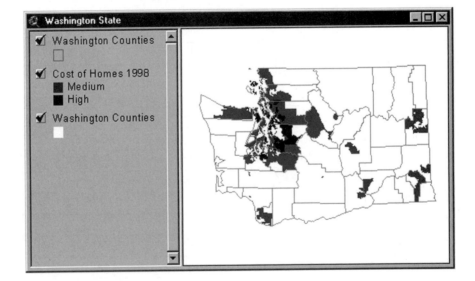

Map scale is also dependent on the scale of the data available for making the map. Always make sure to examine the metadata for information concerning the scale of your data. It does you no good to plan on mapping a feature, such as land ownership, at 1:24,000 scale only to find that the only information available was gathered at 1:100,000 or even 1:500,000 scale. Your map is only as good as the data that goes into it, and gathering new data at large scales can be very expensive and time consuming. An example of a large-scale map is shown in the following illustration.

A large-scale map of Puget Sound in Washington State using a UTM Zone 10 projection.

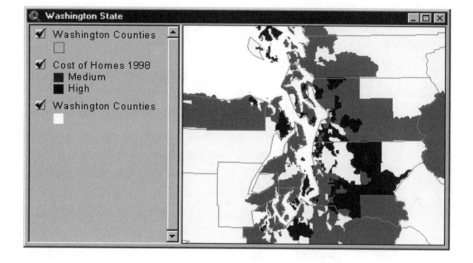

Scale is not necessarily related to detail, although large-scale maps tend to contain more detail. Look at any world atlas. Its small-scale maps will appear crammed with detail, listing hundreds of place names and geographic features. The amount of detail on a map is up to the cartographer, and is determined by the intended use of the map.

Another thing to keep in mind is that scale is not necessarily related to accuracy. Data collected for small-scale maps can be very accurate, whereas some data collected for large scales may be highly inaccurate.

Good metadata will answer many questions about the accuracy of geographic data.

You should use care in mixing radically different scales of data on the same map. Because ArcView allows you to easily add geographic data from a variety of sources in a view (as long as the data is all in the same projection), it is very tempting to mix scaled data. Small-scale data will appear odd at large scales, with rivers or roads showing up as jagged straight lines rather than smooth curves. Conversely, large-scale data can frequently overwhelm a map produced at small scales to the point of illegibility. Before committing to a map project, make sure you have data that has been gathered at similar scales.

Setting Scale in the View

Once you have examined the metadata for themes to determine their scales of use, you can set their appearance at different scales in the view. You can also set the scale for the entire view.

To set the view scale, first set the Map Units in View Properties. Much of the digital data collected today uses meters for map units. Most of the data that comes prepackaged with ArcView uses decimal degrees as map units, whereas some data producers will use feet. Check the metadata of your themes to be sure.

While in the View Properties dialog, you can set the distance units to miles, kilometers, or whatever you like. This will show up when measuring distances in the view and with the scale bar in the layout, but will not affect the scale or projection in the view.

You can set the scale in the Scale window on the far right of the toolbar in the view by typing in a number and pressing <Enter>. The View window will zoom into the requested scale. The Scale window is shown in the following illustration.

The toolbar's Scale window.   Scale 1: 2,483,533   621,692.87 ↔
                                                     5,281,333.98 ↕

Use the Named Extents extension to set reference scales in the view. The Named Extents extension (*namedext.avx*, downloadable from the ESRI ArcScripts web page) allows you to save a specific extent you have zoomed in to, zoomed out from, or panned to in the View window. After loading from the Extensions dialog, a new menu appears called Extents in the View.

Arrange the map display in the View window to reflect the extent you want to show on the map, and select "Add named extent" from the Extents menu. You can save as many extents as you like and return to each by selecting it from the bottom of the Extents menu, all without zooming or panning around the view. The Named Extents menu is shown in the following illustration.

Named Extents menu.

| Add named extent |
| Delete named extent |
| Washington - Puget Sound |
| Washington - Whole State |

You can also set scale for individual themes in the view. Once again, set the map units in the View Properties dialog first, and then make a theme active and go to its Theme properties.

In Theme properties, select the Display icon on the left. The Display panel appears, allowing you to enter minimum and maximum display scales. If you enter *100,000* in the minimum scale box, the theme will be visible down to 1:100,000 scale, but not at 1:50,000 or 1;24,000. If you enter *100,000* in the maximum scale box, the theme will not be drawn in the view until you zoom in to below 1:100,000 in scale, and will appear at 1:50,000 or 1:24,000. The following illustration shows and example of setting a theme's display reference scale.

Setting a theme's display reference scale.

This is a handy means of saving redraw times in the View window when working with a lot of large-scale data.

Projections in ArcView

There are two means of changing projections within basic ArcView, both of which have their pluses and minuses. Projections can be changed on the fly in the View Properties dialog for data stored as decimal degrees (geographic) coordinates, sometimes called unprojected data. Data already stored in a projection will be displayed in the view instantly, as projected, and the projection options in View Properties will have no effect on appearance. In order to permanently change the projection, you will need to use the ArcView Projection Utility extension, or enlist the help of another software program.

Scale and Projection Type Considerations

Whether to store all your geographic data as unprojected or projected depends on how, and to a large extent on where, you work, in that both methods have advantages and limitations. Data stored as decimal degrees gives the user great flexibility in quickly changing projections with the View Properties projection options. You can change perspectives on a geographic area in a matter of seconds. This is of great advantage if you happen to work with data on a worldwide or continental extent. An example of decimal degree data displayed without a projection is shown in the following illustration.

Decimal degree data for North America displayed without a projection.

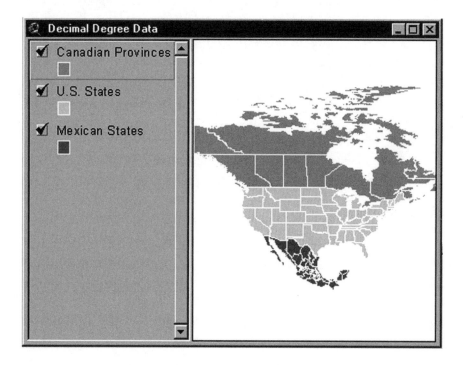

You can change the projection to an equidistant azimuthal type to measure distances between points on the map or to an equal area projection to gauge the size of countries or project areas. You can switch to a world-from-space or orthographic projection and change your global perspective as if you were traveling in orbit aboard the space shuttle. This is particularly valuable for educators or anyone concerned with global issues, such as population growth or environmental change.

Many, if not most, ArcView users work in much smaller geographic areas, and may be limited to a single state, province, or city. At these larger scales, the ability to change projections quickly through the View Properties dialog is much less useful and can even be detrimental. People working with data at less than approximately 1:500,000 scale for a single area (Seattle and the Puget Sound area, for example) will often keep all of the their information in a single projection such as Washington State Plane North or UTM Zone 10. The projection will be chosen to minimize distortion of areas and distances, and allow the user to make accurate depictions of feature locations.

At scales greater than 1:100,000, such as 1;24,000 USGS topographic scales, the "on the fly" projection feature in View Properties can introduce significant error to the location of map features. Point features carefully collected with a GPS unit, differentially corrected, and then projected on the fly in ArcView may be anywhere from several hundred meters to several kilometers from their true locations. A good test is to project the same data using the ArcView Projection Utility, ArcInfo, or the GPS software and then overlay it with decimal degree data projected on the fly and then saved in the projection using the Convert to Shapefile option from the Theme menu.

This discrepancy happens because the projections in View Properties are using different algorithms to recalculate the location of features than the projections produced with the ArcView Projection Utility. The projections in the ArcView Projection Utility take much longer to calculate, but result in more accurate locations. Because the same projection engine is used in the ArcView Projection Utility, ArcInfo

8.0 and MapObjects 2.0, all three ESRI software packages should produce the same accurate projections.

✓ **TIP:** *When locational accuracy is important at large scales, always use data that has been projected in something other than the View Properties projection feature.*

Projection Files and Metadata

A crucial step in changing projections is knowing what projection your data is in to start with. With shapefiles, you either have projection information (along with scale information) included with the theme's metadata or the shapefile comes with an additional projection file.

Projection files have a *.prj* suffix and are new to ArcView 3.2. The projection file is a simple text file that can be opened with any word processor, containing the projection type, map units, and projection parameters. Projection files are generated for shapefiles anytime you use the ArcView Projection extension, and should be stored in the same directory as the parts of the shapefile.

These new files will work in ArcView 3.2, ArcInfo 8.0, and ESRI's MapObjects 2.0, allowing projection information to be interchangeable across the ESRI product line. When opened, projection files appear all on one line of text, but have been broken up in tables 7-2 and 7-3, which follow, for ease of reading. Table 7-2 lists ArcView projection files in decimal degrees for the continental United States. Table 7-3 lists ArcView projection files in Albers Equal Area projection for the continental United States.

Table 7-2: ArcView Decimal-Degree Projection Files for the Continental United States

Item Name	Data
GEOGCS	"GCS_North_American_1983"
DATUM	"D_North_American_1983"
SPHEROID	"GRS_1980",6378137,298.257222101
PRIMEM	"Greenwich",0
UNIT	"Degree",0.0174532925199433

Table 7-3: ArcView Projection Files in Albers Equal Area Projection for the Continental United States

Item Name	Data
PROJCS	"Custom"
GEOGCS	"GCS_North_American_1983"
DATUM	"D_North_American_1983"
SPHEROID	"GRS_1980",6378137,298.257222101
PRIMEM	"Greenwich",0
UNIT	"Degree",0.0174532925199433
PROJECTION	"Albers"
PARAMETER	"False_Easting",0
PARAMETER	"False_Northing",0
PARAMETER	"Central_Meridian",96
PARAMETER	"Standard_Parallel_1",29.5
PARAMETER	"Standard_Parallel_2",45.5
PARAMETER	"Central_Parallel",37.5
UNIT	"Meter",1

Without a projection file or metadata, you are reduced to guessing what projection the original data is in. This is never a good move. Try looking at the coordinates shown to the right of the scale gauge on the toolbar in the View window and comparing them to another view with data in a known projection. This sometimes works, but can also backfire, in that the differences between some projections can be quite subtle. The lesson is that nothing makes up for good metadata!

Changing Projections in View Properties

ArcView will change projections of data in geographic coordinates (in decimal degrees of latitude and longitude) on the fly within a view window. The data must be in the ArcView shapefile format, and all data within a view window will be changed at once. That is, you cannot have two themes in different projections in the same view. Neither can you re-project raster data or ArcInfo coverages in this manner.

When you open the View Properties dialog, you will notice that ArcView has detected that the data is in geographic coordinates, and that the map units have been automatically filled in as decimal degrees. The Projection should appear as "None," indicating that you are able to project the data from the Projection button. Click on the Projection button to bring up the Projection properties, which will show that the data is in a standard geographic projection. The View Properties projection feature is shown in the following illustration.

*View Properties
projection feature.*

There are eight categories of standard projections available, as well as twenty custom projections. With a standard projection, the parameters (such as the central meridian) reference latitude or standard parallels, are hard wired, and are not selectable options, as they are with custom projections.

> ✓ **TIP:** *After you have projected the themes in a view with the View Properties projection feature, you can save active themes in that projection by using the Convert to Shapefile option in the Theme menu. You will be prompted to rename and save the active theme, and whether to add it to another view. This is a quick method of permanently re-projecting themes from decimal degrees to another coordinate system.*

The following lists are of standard projection categories and the projection options (or descriptive information concerning the options) available under them in the View Properties dialog.

- Projections of the World

  - Behmann
  - Equal-Area Cylindrical
  - Geographic
  - Hammer Aitoff
  - Mercator
  - Miller Cylindrical
  - Mollweide
  - Peters
  - Plate Carree
  - Robinson
  - Sinusoidal
  - The World from Space

- Projections of a Hemisphere

  - Equidistant Azimuthal (Equatorial)
  - Equidistant Azimuthal (North Pole)
  - Equidistant Azimuthal (South Pole)
  - Gnomic (Equatorial)
  - Gnomic (North Pole)
  - Gnomic (South Pole)
  - Lambert Equal-Area Azimuthal (Equatorial)
  - Lambert Equal-Area Azimuthal (North Pole)
  - Lambert Equal-Area Azimuthal (South Pole)
  - Orthographic (Equatorial)
  - Orthographic (North Pole)
  - Orthographic (South Pole)
  - Stereographic (Equatorial)
  - Stereographic (North Pole)
  - Stereographic (South Pole)

- Projections of the United States

  - Albers Equal-Area (Alaska)
  - Albers Equal-Area (Coterminous U.S.)
  - Albers Equal-Area (Hawaii)
  - Albers Equal-Area (North America)
  - Equidistant Conic (Coterminous U.S.)
  - Equidistant Conic (North America)
  - Lambert Conformal Conic (Coterminous U.S.)
  - Lambert Conformal Conic (North America)

- State Plane – 1927

One hundred and twenty sections for the United States based on the North American Datum of 1927, which uses Clarke's 1886 spheroid.

- State Plane – 1983

One hundred and twenty sections for the United States based on the North American Datum of 1983, which uses the GRS80 spheroid.

- UTM – 1927

Sixty 6-degree-wide zones based on the North American Datum of 1927, which uses Clarke's 1886 spheroid.

- UTM – 1983

Sixty 6-degree-wide zones based on the North American Datum of 1983, which uses the GRS80 spheroid.

- National Grids

  - Great Britain
  - New Zealand

- Malaysia and Singapore
- Brunei

Projections of the World

The standard projections offer twelve projections that will show the entire world in the View window. These world projections work best when you have the data for the entire world, such as the outlines of continents, population by country, ecoregions of the world, and so on. Some are quite specialized and rarely used; others you may see commonly on printed maps, such as the Mercator or Robinson projections. The World from Space projection is quite useful for getting a feel for the global relationships of geographic data.

Projections of a Hemisphere

There are fifteen hemispherical projections available, broken up into five types. Each type offers a north and south polar projection and an equatorial projection. The standard equatorial projection type is based on the Western Hemisphere, with a central meridian of −72.53 degrees. This meridian lies just off the east coast of North America.

In order to see the Eastern Hemisphere, click on the Custom radio button and remove the negative sign. The view will appear to rotate the globe, and center the Eurasian landmass in the window. In fact, keep the custom projection turned on and you can continue to rotate Earth by whatever increment of longitude you wish.

➥ **NOTE:** *ArcView treats everything west of the prime meridian at Greenwich as a negative number, and everything east as a positive number. 180 degrees west or east is roughly the international dateline in the central Pacific Ocean.*

Projections of the United States

As you might expect, no one projection can accurately map a country such as the United States, stretching from Maine to Hawaii. Therefore, ArcView provides eight standard projections of the United States, broken up into types that focus on the coterminous United States, Hawaii, Alaska, and the rest of North America. The Albers and Lambert projections included here are also probably the best bets for showing the large expanse of Canadian provinces and territories.

State Plane Projections

The State Plane Coordinate system was adopted in the 1930s to establish a grid of constant scale throughout the United States. Each of the 48 states at that time was broken up into a series of either vertical or horizontal zones, depending on the orientation of the individual state. States wider east to west were projected using a Lambert Conformal projection, whereas states taller north to south were mapped using a Transverse Mercator projection. At the time, the most advanced datum was the North American Datum of 1927, which uses Clarke's 1886 spheroid, hence the name State Plane NAD27. ArcView has all 120 of these historical projections available.

Although many historical maps use this older coordinate system, it was significantly revised in the 1980s. Many states eliminated the cumbersome multiple-zone system, and adopted a single-zone system. And of course, there were two new states. The Alaska panhandle presented a special problem, being aligned neither north to south nor east to west, but diagonally from the northwest to the southeast. An oblique Mercator projection was adopted for this portion of the state.

The North American datum had finally been updated as well, to NAD83. ArcView contains all 125 or more sections of these new state plane projections. The new projections are read, using Montana as an example, as Montana State Plane Single Zone NAD83.

UTM Projections

The Universal Transverse Mercator projections (UTMs) are popular because of their ease of use worldwide. Earth is divided into 60 6-degree-wide numbered UTM zones starting at 180 degrees west longitude. The coterminous United States is divided up into 10 UTM zones, starting with UTM Zone 10 in California and proceeding to UTM Zone 19 in Maine. UTM zones for the United States are shown in the following illustration.

UTM projections are very useful for large-scale maps, such as the USGS 7.5-minute topographic map series, on which UTM coordinates divide the map into squares 10,000 meters on a side. A UTM projection not only minimizes distortion at this large scale but makes it easy to read UTM coordinates directly from a paper map.

UTM zones for the United States.

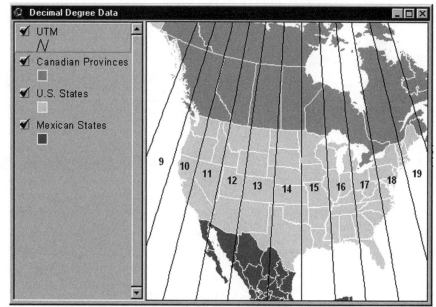

One disadvantage of UTM projections for states with wide east-to-west extents (or any area for that matter) is that several zones may be needed to cover a state, making it impossible to fit the entire area in one UTM projection without distorting the edges. ArcView offers UTM projections with the older NAD27 datum and the more current NAD83 datum.

National Grids

Four national grids are available in the View Properties projections for Great Britain, New Zealand, Malaysia/Singapore, and Brunei. The Canadian grid systems are not supported yet in either View Properties or the ArcView Projection Utility.

Custom Projections in the View Properties

Twenty custom projections are offered in the View Properties dialog, in which you have the opportunity to select your own parameters for a projection. For those new to map projections, this is not as difficult as it seems. The View Properties custom projections and parameters are shown in the following illustration.

Custom projections and parameters in View Properties.

Projection Properties	
○ Standard ⊙ Custom	OK
	Cancel
Projection: Lambert Conformal Conic	
Spheroid: Clarke 1866	
Central Meridian:	-115.4189397909
Reference Latitude:	46.4347621835
Standard Parallel 1:	21.9857222101
Standard Parallel 2:	70.8838021569
False Easting:	0
False Northing:	0

First, once you choose a custom projection, ArcView will guess at appropriate parameters. You can use these as a starting point for your own by altering one parameter at a time and appraising the result in the View window. Alternatively, select a standard map projection to start with, and click on the Custom radio button, which will allow you to alter the projection.

The number of parameters that need to be entered range from two to seven, depending on the type of custom projection selected. Before becoming too committed to a certain projection, look up specifics on that projection in the ArcView Help files. The following are the custom projections available in the View Properties dialog.

- Albers Equal-Area Conic

- Cassini Soldner

- Equal-Area Cylindrical

- Equidistant Azimuthal

- Equidistant Conic

- Equidistant Cylindrical

- Gnomic

- Hammer-Aitoff

- Hotine Oblique Mercator

- Lambert Conformal Conic

- Lambert Equal-Area Azimuthal

- Miller Cylindrical

- Mercator

- Mollweide

- Orthographic

- Robinson

- Sinusoidal

- Stereographic

- Transverse Mercator

- Vertical Near-Side Perspective

Using the ArcView Projection Utility

ArcView 3.2 introduced the ArcView Projection Utility extension, based on the new projection engine for ESRI products. The ArcView Projection Utility replaces the old Projector! extension, which was based on the View Properties projection feature.

The new utility is meant to provide the same accurate projections for all ESRI software products. It also requires a knowledge of projection terminology, some of which has already been covered, and an increased level of patience, as you wait for the wizard-based utility to do its thing. As with the View Properties projection feature, this incarnation of the Projection Utility only works with ArcView shape-files, and not with ArcInfo coverages or raster data such as images or DEMs.

Getting Started with the Projection Utility

The ArcView Projection Utility changes coordinate systems, projections, and datums, but can be intimidating to the new user. It behooves the user to have some knowledge of projection terminology, as well as metadata on the themes to be projected.

✓ **TIP:** *Have all shapefiles (with the same projection) you want to change in the same directory. A batch process for changing projections is much faster than doing it one at a time. Also, work out in advance what projection you want to change to. You can waste a lot of time experimenting with this slow utility.*

The utility is automatically placed in the ArcView extensions folder when ArcView version 3.2 is installed, and is therefore loadable from the Extensions dialog. You begin to open the wizard-based utility after selecting ArcView Projection Utility in the File menu in the view. Then you wait.

✗ **WARNING:** *The initial introduction panel takes anywhere from one minute to five minutes to open, depending on the speed of the computer's CPU. Have patience, and do not use <Control-Alt-Delete> thinking that your machine has frozen! Click on the Next button when it finally appears.*

Step 1: Select the Shapefiles to Be Projected

The first number panel appears, as shown in the following illustration. You can browse directories to find the shapefiles you want to project, and select multiple files by using the <Shift> key.

Step 1 in the ArcView Projection Utility.

ArcView Projection Utility - Step 1

Select which shapefile(s) you would like to reproject into another coordinate system. If multiple files are specified, they must exist in the same directory and be in the same coordinate system.

Directory
d:\gisdata\projections\dd   Browse...

Name	Size	Type	Count	Coordinate System
states.shp	218KB	POLYGON	51	GCS_North_American_1983
roads.shp	247KB	LINE	679	GCS_North_American_1983
rivers.shp	105KB	LINE	56	GCS_North_American_1983

3 file(s)

Help   Cancel   < Back   Next >

The panel shows the names of the shapefiles, their size in kilobytes, their type, and a count of features in each file. Shapefiles with a lot of features will take longer to process.

If the shapefiles already have a projection file (.*prj* file), the utility will show that coordinate system in the far right column. Otherwise, the utility will list the coordinate system as unknown.

> ✓ **TIP:** *Click with the right mouse button in the directory window of this panel to see additional projection information, if available, for the selected shapefiles.*

> ↝ **NOTE:** *All shapefile data that comes with the ArcView program will already have .prj files.*

Once you have selected the shapefiles you want to work with, click on Next.

Step 2: Select the Current Projection of the Shapefiles

If no projection file exists, the panel for step 2 will ask for the current coordinate system of your shapefiles. Without good metadata here, you are going to spend a lot of time guessing!

Click on the box in the upper right of the panel to show advanced options. This will bring up tabs for projection parameters, datum, and ellipsoid.

Your first choice in the Name tab, shown in the following illustration, is whether the data is in a geographic or projected coordinate system. A shapefile in geographic coordinate system has data stored as latitude and longitude in decimal degrees. This describes most of the data that comes from ESRI with ArcView. A shapefile in a projected coordinate system will have data stored in *x,y* fashion, and will already be projected. Select either Geographic or Projected, depending on the metadata information you have for the themes.

Step 2, using the Name tab in the ArcView Projection Utility.

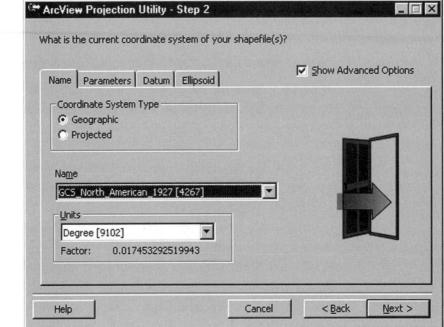

There are several dozen choices for geographic coordinate systems, under the Name drop-down box, and they have a different naming convention than ArcView users are used to. Take the following choice, for example: *GCS_North_American_1983[4269]*. This is read as the Global Coordinate System, being used in North America, with the NAD83 datum. The last number in brackets is called the POSC Code Number. The ArcView Projection Utility uses projection algorithms based on an international standard developed by the Petrotechnical Open Software Corporation (POSC), and each projection has a unique coded number that can be used to call that projection with Avenue.

There are several hundred choices for projected coordinate systems, and it can be quite a task to find the one you want in the cramped drop-down box. The following are examples of a couple of common projected coordinate systems for North America.

- *NAD_83_Montana[32100]:* Indicates the new state plane projection for Montana, with POSC number. All state plane projections will begin with the datum, NAD83 or NAD27, followed by the state's name.

- *NAD_1983_UTM_Zone_12N[26912]:* Indicates a UTM Zone 12 projection, north of the equator, followed by the POSC number. All UTM projections will start with the datum, followed by the UTM zone number.

➤ **NOTE:** *For help with coordinate systems for countries outside North America, examine the ArcView help files for the projection utility.*

Below the coordinate system type is a box for units. Data in a geographic coordinate system will probably have degrees for units. Data in a projected coordinate system is usually in meters or feet (check the metadata again!), but other peculiar units are offered, including yards, chains, links, and fathoms.

Click on the Parameters tab next, and examine the parameters. If you selected an existing projected coordinate system on the Name tab, the utility will fill in standard parameters such as the prime meridian, false northing and easting, and the coordinate system's base projection. If you select Custom on the Name tab, you will be prompted to enter custom parameters for the projection of your choice. The Parameters tab is shown in the first of the following illustrations.

Click on the Datum tab next, which is shown in the second of the following illustrations. Existing coordinate systems will have the datum information filled in; custom systems will not.

*Step 2, using
the Parameters
tab in the
ArcView
Projection
Utility.*

✓ **TIP:** *The Datum tab is where you need to set a geographic transformation if you are changing datums in your projection (between NAD27 and NAD83, for example). There are a couple of dozen geographic transformations to select from in the drop-down box in the upper right of the Datum tab. If you cannot find the one that matches your transformation, try transforming the existing datum to the World Geodetic System 1984 (WGS_1984). Complete the steps in the projection process, and run it again to take your shapefile from WGS_1984 to your new datum.*

Step 2, using the Datum tab in the ArcView Projection Utility.

Last, click on the Ellipsoid tab, shown in the following illustration. It should already be filled in, unless you have chosen a custom coordinate system.

Click on the Next button. If your shapefiles did not come with a *.prj* file, the utility will prompt you to save the last tabs of projection information. It is always good to do this, because saving *.prj* files now will save you time in the future, and will provide the next user of those shapefiles with another piece of metadata.

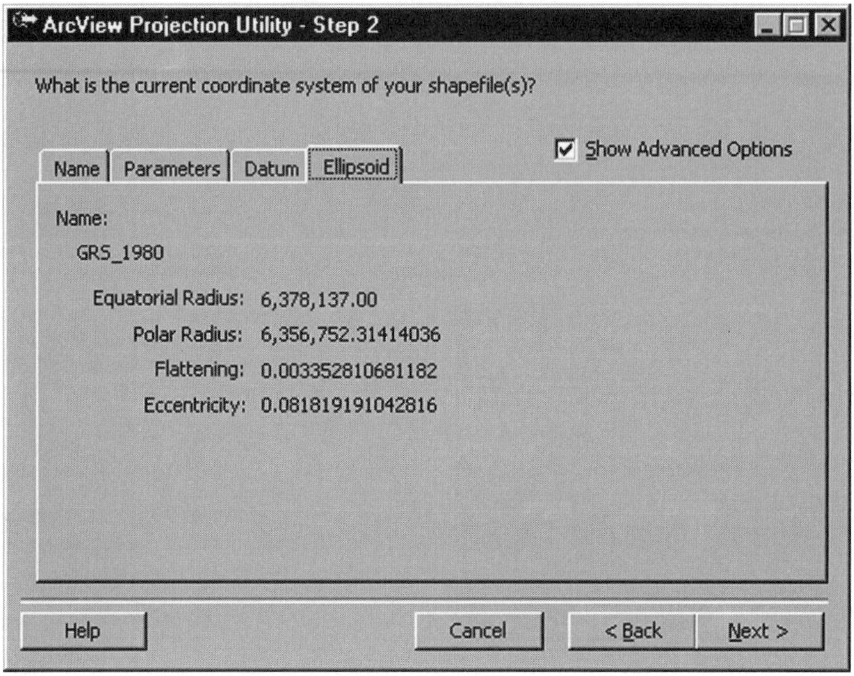

*Step 2, using
the Ellipsoid
tab in the
ArcView
Projection
Utility.*

Step 3: Select a New Coordinate System for the Shapefiles

Panel 3 asks you to define the new coordinate system for your shape-files. Follow the same procedure as described for step 2, going through the four tabs, and click on Next.

Step 4: Select Where to Save the New Shapefiles

The utility will ask you where to save the new projected shapefiles. This is shown in the following illustration. Choose a directory and click on Next.

Step 4 in the ArcView Projection Utility.

✓ **TIP:** *Save the new projected shapefiles in another directory to avoid confusion with the old shapefiles. It is wise to keep all shapefiles in a single directory in the same projection.*

Step 5: Summary Information

The utility in step 5 will present a summary of the information you have selected in the previous three steps, as shown in the following illustration. Review the projection information, and use the Back button to change any mistakes. If the information looks good and you intend to use this tiresome process again, use the Print button to produce a hardcopy description. Put the paper in a safe place for future reference.

*The summary step
in the ArcView
Projection Utility.*

Click on the Finish button, and go get a cup of coffee, if not something stronger. The ArcView Projection Utility will grind away, even if you exit ArcView at this point, taking anywhere from a couple of minutes to a half hour to re-project the shapefiles. The utility will send a message when it is done.

Alternatives to the Projection Utility

Once you have figured out the ArcView Projection Utility and its associated jargon, using it is actually not too bad. However, it does no process quickly. If you have to re-project data frequently, and need quick, accurate projections, there are several options. You may be lucky to have access to workstation ArcInfo in your office, which will perform the same projections much faster. Arc will also handle GRID files the ArcView Projection Utility does not. The ESRI Data Acquisition Kit, called DAK for short, has a good set of projection tools taken from PC ArcInfo, but it will cost you as much as another copy of ArcView.

If you are using data gathered with GPS units, check out the projection possibilities inside the software that may have come with your GPS setup. Trimble Pathfinder software does a fine job of re-projecting GPS data and exporting it as a finished ArcView-compatible shapefile.

For those limited to ArcView, Blue Marble Geographics offers two fine ArcView extensions that can speed up the projection process. ShapeProjector AVX (retailing at $199) will do the same functions as the ArcView Projection Utility, but a heck of a lot faster. ShapeProjector's AVX dialog is shown in the following illustration. Geographic Transformer AVX (selling for $499) will do the same for image files, something you will not find outside high-end image processing GIS systems. Both are available on the web at *http://www.bluemarblegeo.com*.

ShapeProjector AVX dialog.

Summary

Projections are a powerful, but complicated, tool for transforming the curved surface of Earth onto a flat map. Cartographers using ArcView need to decide the extent and scale of the area they want to map, and select a projection that meets the map's purpose, without misleading their audience.

Chapter 8

Cartographic Design with the Layout Document

The ArcView layout document is where you put the finishing touches on your cartographic product. It is here you set the size of the map product, and finally assemble the five parts of a good map. You have already done most of the heavy lifting of map construction in the view document, with the Legend Editor and the feature labeling tools. Now it is time to add the design touches that produce a superior cartographic product.

When ArcView version 2.0 introduced the layout document, it revolutionized the speed at which complicated GIS data could be placed on an attractively designed map. The layout document presented the concept of live links, where changes in the view would automatically appear in the map and legend on the page.

The ArcView layout provides a series of tools called frames, which place the map display, the map legend and scale, the compass or North arrow, and allow the addition of charts and tables to the page. The frame tools, in conjunction with a several other graphic tools, give you the capability of adjusting visual balance, contrast, and the cartographic order of map elements.

Table 8-1: ArcView Layout Process Checklist

Location and Tool	Task
In the AV layout document:	
Layout Properties:	Create new layout document and choose page size
Frame Tools:	Place view frame on page and size to view and legend frame on page and size to fit
On Paper:	Sketch draft map layout
	Position major map elements
In the AV Layout Document:	Adjust layout document
Frame Tools:	
• View Frame	Adjust view frame
• Legend Frame	Place map legend
• Text Tool	Place map title
• Scale Frame	Place scale bar
• North Arrow Frame	Place North arrow, if needed
• Text Tool	Place map title
• Legend Frame	Simplify legend, if needed
• Text Tool	Place source and copyright information
• Neatline Button	Draw neatlines
On the Printer:	Print draft on printer
In the AV Layout Document:	Revise layout and print final map

Design Considerations

Design considerations include such things as page size, composition, type style, and numerous other elements that affect the aesthetic and functional result of a map. These considerations are discussed in the sections that follow.

Page Size

Although the layout document is truly the graphic design arena, and not a geographic analysis tool like the view, certain actions undertaken here control other actions in the view. The choice of page size for your map, for example, acts as a control of the map scale, the amount of geographic detail, and the amount of feature labeling you can do in the view document. That is why determining the final map size, at least approximately, should be one of the first decisions made when undertaking a map project.

The page size also determines the amount of room for your map legend. A land cover map with a hundred classes of vegetation will not have room for a map legend on something smaller than letter size.

Map Composition

Arranging the elements of a good map in an attractive and readable manner is an art that takes patience, experience, and a good eye for design. If you are new to mapping and lack design experience, look at other examples of maps to get ideas on how elements are arranged, and try to imitate the good ones.

A great place to start is the ESRI Map Book, published annually by Environmental Systems Research Institute. The Map Book, which can be ordered on line at *www.esri.com,* features about 100 of the best maps produced by users of ESRI software products, including Arc-View, and can stimulate dozens of ideas on how to present your map.

Major and Minor Elements

You can start by thinking of the map elements as major or minor pieces, and experiment by moving them around graphically on the screen, until you reach a design that fits for that particular map. The major elements are the map display itself, the legend, and the map title. These elements invariably take up the most room on your page, and should be the most noticeable and quickly readable.

The minor elements are the map scale and projection, a North arrow or compass, a source statement, and any other text. These elements, though important, should not overpower or dominate the major elements. Remember that your map purpose is conveyed by the map display itself and reinforced by the map title and legend.

Map Scale and Map Elements

One convention concerning the arrangement of map elements holds that with large-scale maps, such as 1:24,000 scale topographic maps, additional map elements are placed in the margins outside the map display, so as not to obscure map details. Examples of small- and large-scale map layouts are shown in the following illustration.

Small- and large-scale map layouts.

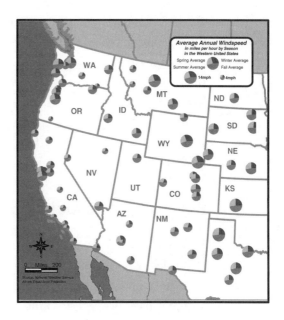

Small-scale map

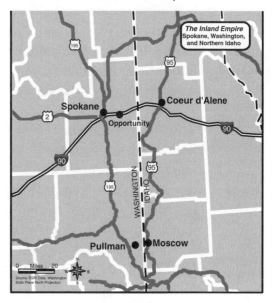

Large-scale map

With small-scale maps, additional map elements may be located within and actually overprint the map display. Elements such as legends or map scales may be placed over portions of the map display that have little information, or that are unimportant to the overall map purpose. It is up to the cartographer to decide whether these elements are legible overprinting the underlying map display or require being placed in a box, filled with a color to block out the underlying map detail.

Balance and Readability

When arranging the map elements on the page, there are no hard and fast rules. You should, however, strive for readability and attempt to achieve a visual balance of map elements that is pleasing to the eye, and that basically "looks right."

Consider the flow of the reader's eye as it passes over the map on the page. In the Western world, people read from top to bottom, and from left to right on the page. As a result, a map title is commonly near the top of the page, whereas minor elements, such as the source statement or map scale, may be found near the bottom.

Consider also the visual center of the page, which readability studies show is a point slightly above the actual center of the page. The most important features shown on your map display should appear at or near this visual center, with additional map elements placed around it in a visually balanced fashion. Examples of map element positions are shown in the facing illustration.

Depending on the shape and projection of features on your map display, the arrangement of the other map elements may be easy or quite difficult. A small-scale map showing the continental United States has a lot of water area for placing legends, but this might not be the case with a large-scale urban map of a city such as Denver.

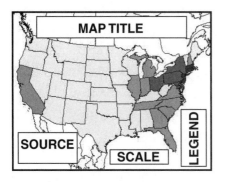

Small-scale map:
elements placed on map.

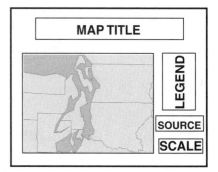

Large-scale map:
elements placed around map.

Combinations of different positions for map elements.

It is always important to avoid crowding the map with too many details, or with large legends in boxes, large blocks of margin text, or giant North arrows. Use the power of negative space or white space carefully to increase readability, without leaving large gaps or holes devoid of information in the map composition. When arranged properly, the map composition should enhance figure-ground contrast between the map's central purpose and the other elements in the layout.

Additional Elements and the Use of Neatlines

ArcView allows you to go beyond the basics in map composition centered around a single map display and add multiple displays of other views. Other views can act as locator maps for or close-ups of the main map display, or portray a times series of data for the area. Your creativity here is only limited by your data!

Blocks of explanatory text can be added, turning the map into a storyboard or an info-graphic. All good maps tell a story, and that story can be enhanced with the judicious use of text, describing what the map is showing and its significance to the reader.

As more elements such as view frames and text blocks are added to the composition, it may become necessary to add neatlines or graphic boxes to separate the elements from the main map display. Care should be taken to avoid the overuse of neatlines, which tend to result in extremely boxy page layouts that detract from the map.

Organizational Style

If you are creating a series of maps for a project or report, it is good to adopt a similar organizational style for all of the map products. A similar arrangement of map elements for each map product improves the readability of the entire map series, and helps the map reader quickly understand the purpose of each composition. Good map compositions can be stored by ArcView as templates and retrieved for later use.

Type Style for Map Elements

By the time you have reached the layout stage, you have already labeled all map features you need in the view document, with the labeling tools. In contrast to the map display itself, which can use from one to three font families, the rest of the margin information should be in one font family, including the legend. Use bold sparingly, possibly just for the title, as to not draw attention away from the map.

Legend Style

ArcView by default automatically places all the themes visible in the view in the map legend. This feature can result in long, complicated legends with obvious map features highlighted (such as roads, rivers, and lakes) to the detriment of the major themes. Use the Simplify feature to break a complex legend into its constituent parts and delete themes that are obvious to the reader.

Leaving just the relevant themes in the legend increases the readability of the map composition and ensures that you will get your point across to your intended audience. Shorter legends also increase the likelihood that they can be placed to fit over the map display in a small area, and will not require a large opaque box behind them to make them legible. Legend styles are shown in the following illustration.

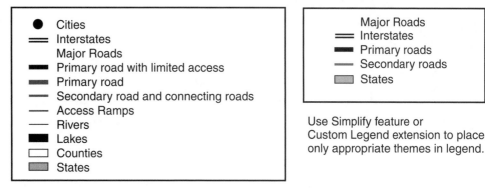

Default ArcView legends include all active themes.

Use Simplify feature or Custom Legend extension to place only appropriate themes in legend.

Legend styles.

Because tweaking the map display in the view in order to change its appearance in the layout is a common task throughout the mapmaking process, the legend should be the last item finished on the page. Wait until the last stages of the process to use the Simplify feature to shorten your legend.

When to Add a North Arrow or Compass

In the same manner as deleting obvious themes from a legend, a North arrow is unnecessary on many maps where the orientation of the map is obvious to the reader. Small-scale maps showing entire continents or containing latitude-longitude grids do not need a North arrow. Large-scale maps with a lot of detailed local features the reader may not be familiar with need a compass, as well as maps where north is not at the top of a page.

ArcView's North arrows tend to be large and noticeable, and can take up a lot of room on the page. Try making the North arrows smaller and not much larger than the height of the map's scale bar.

Using the Page Setup and Layout Properties

From the start of your mapping project, when you began placing themes in the view document, you have had a final page size in mind for the map product. Before beginning to place your map elements on the page, you need to choose the size, orientation, and margin of that page in the Page Setup dialog, shown in the following illustration.

Page Setup dialog box.

Go to the Layout menu and select Page Setup. In the Page Size drop-down menu, select the page size you need for the map. The default feature, "Same as printer," will usually be letter size for most printers. You can select English or metric measurements in the Unit drop-down menu next, and then choose an orientation, either landscape (wide) or portrait (tall). Table 8-2, which follows, lists common page sizes used in the Page Setup dialog.

Set margins by either using the printer's default border, or setting a custom margin size. Last, select an output resolution for raster images, if any, in your layout. For more discussion about printer margins and output resolution see "Margin Size and Printing Time per Copy" in Chapter 11.

Table 8-2: Common Page Sizes Used in the Page Setup Dialog

Name	*Size*
Same as printer	Printer default size
Letter	8.5 x 11.0 in
Legal	8.5 x 17.0 in
Tabloid	11.0 x 17.0 in
C	17.0 x 22.0 in
D	22.0 x 34.0 in
E	34.0 x 44.0 in
A4	21.0 x 29.7 cm
A3	29.7 x 42.0 cm
A2	42.0 x 59.4 cm
A1	59.4 x 84.1 cm
A0	84.1 x 118.9 cm
Camera	11.0 x 7.33 in
Custom	User-specified size

To change page sizes easily for your map product use the Template feature in the Layout menu. To change a letter size to a larger poster size, such as E size, first finish a map in the layout at the letter size, being careful to place and arrange all map elements exactly as you want them. Next, use the Store As Template feature in the Layout menu to save your letter-size map as a custom template, with a unique name.

Now you are ready to set up another layout document with a larger page size. In the AV application window, select New to create another layout document. In the new layout, go to Page Setup and select E Size (34.0 x 44.0 inches). Make sure the page's orientation is the same as your original letter-size map. Finally, go to Use Template in the Layout menu, and select your custom template. The E-size page is filled with the map from the smaller letter-size product.

The Layout Properties dialog, shown in the following illustration, in the Layout menu allows you to name your layout and manipulate the background grid on the page. It is always a good idea to name each layout something other than "layout1, layout2," and so on. A nice naming trick is to include the page size with each name, such as "Wind Speed Map—Letter Size" or "Wind Speed Map—E Size." This avoids confusion and adds clarity when selecting existing layouts from the list in the ArcView application window.

Layout Properties dialog box.

The Snap to Grid feature in the Layout Properties box is a means of organizing map elements on the page. However, this tool is frequently more irritating than useful. Supposedly, map elements should easily snap to vertical and horizontal grid lines on the page, but instead most users have trouble dragging and aligning the elements correctly. Check this feature off in Layout Properties, and use the Graphic Size and Position, Align, and Group features in the Graphics menu instead.

Using the Frame Tools and Properties

All of the map elements, except for the map title and source statement, are placed on the page with the use of the ArcView frame tools. Located on a drop-down icon bar on the toolbar in the layout document, the frame tools control the size, placement, and behavior of the map element. There are seven frame tools, listed in the following illustration. Four tools control major map elements, and three additional tools place charts, tables, and pictures.

Frame tools.

Creates a view frame on the display

Creates a legend frame on the display

Creates a scale bar frame on the display

Creates a North arrow on the display

Creates a chart frame on the display

Creates a table frame on the display

Creates a picture frame on the display

The first three frame tools (the View, Legend, and Scale Bar Frame tools) are *view dependent.* That is, they have to be linked to the same view document for the resulting map to make sense.

All of the frame tools behave in the same manner. You click on the tool icon in the drop-down icon bar, go to the page in the layout window, and define a box by click-dragging from the upper left to lower right. A Frame Properties box appears, in which you set a variety of options, and then click on OK. The map element appears on the page.

The resulting frame box can be moved by selecting it with the Pointer tool (the four black graphic handles should appear), and dragging it to a new location on the page. Frame boxes can be resized by selecting them with the Pointer tool, clicking and holding down the mouse on one of the graphic handles, and dragging diagonally in or out. ArcView, unlike other graphic programs, will automatically constrain the resized graphic element, so that the graphic does not appear stretched or misproportioned.

You can reaccess the frame's properties dialog box at any time by double clicking in the center of the frame box. The frame properties dialog reappears, allowing you to change any of its options, such as basing the frame on a different view document entirely.

The View Frame

The View Frame tool is the primary tool for placing your map on the page. The tool takes information as shown in the map display portion of the view document and places it in the drawn rectangle on the page. As the primary tool for placing map information in the layout document, it has the most number of options in its properties dialog box, as shown in the following illustration.

View Frame Properties dialog box.

View Frame Properties

View: View2
Wind Speed

☑ Live Link

Scale: Automatic

1: 0

Extent: Fill View Frame

Display: When Active

Quality: Presentation

OK   Cancel

Selecting a View

After dragging a frame box on the page and releasing the mouse, the View Frame Properties dialog opens. You are asked which view window to base the frame on. The view frame can be "live linked" to the view document; that is, as the view changes with themes turned on or off, or re-symbolized, the changes appear on your layout page in the view frame. This innovation, introduced in ArcView 2.0, essentially changed the way desktop mapping is done, allowing changes to be seamlessly integrated in your map product.

Live Links

Although live link checked on is the most common way of placing a view window, you can turn the feature off, and ArcView will "remember" the view document status as of the time you placed it in the layout, no matter how many changes transpire in the view. This is one way of producing a series of closely related maps, with several layouts based on one view document, and incremental changes made in the view reflected in the layout with live link checked off. However, it can become difficult or impossible to go back to the original view document and tweak earlier layouts in the map series, and Arc-View may "forget" what an older view window was displaying.

> ✓ **TIP**: *A better way of producing a series of closely related maps is to use the ArcView ODB extension version 1.2 to clone your layouts and views. See the discussion on using the ArcView ODB extension in Chapter 9.*

Map Scale

There are three options for setting map scale in the layout. The default option is Automatic, which takes the width of the view's display window and fits that width in the drawn view frame on the page. This of course changes the scale of the layout to fit the width of the view display.

You can also select Preserve View Scale, which will cause the view frame to use the same scale as the view document, even if the view frame is too small to contain all of the map display in the view. If live link is checked on, the view frame's scale will change if the view's map display scale is changed. User Specified Scale is the last scale option, and requires you to type in a numeric scale.

> ✓ **TIP**: *As one of the first steps in any ArcView mapping project, set the Map Units and Distance Units in the View Properties dialog box in the view. Set map units are required in order to use the User Specified Scale option in the view frame properties dialog box, and distance units are needed to create a usable map scale in the Scale Frame Properties box.*

Map Extent

You can either Fill View Frame or Clip to View for a printed map extent in the layout. Fill View Frame will sometimes take extra data beyond the edges of your view window and make it visible in the layout if the view frame is larger than the map display window in the view. Clip to View will limit the data shown in the view frame to the exact extent of the map display in the view window, no matter how large the view frame is drawn. Table 8-3, which follows, presents combinations of map scale and map extent properties.

Table 8-3: Combinations of Map Scale and Map Extent in View Frame Properties

Scale	Fill View Frame	Clip to View
Automatic	View fits in view frame. Extra data fills in frame.	View fits in view frame. White space fills in around the centered view in the frame.
Preserve View Scale	View and view frame are the same scale. If frame is too small, portions of view will not appear.	View and view frame are the same scale. If the view frame is too large, white space will fill in around the view in the frame.
User Specified Scale	View frame scale is fixed, and not dependent on the view scale. If the view frame is smaller than the view, only a portion of the view will appear in the view frame.	View frame scale is fixed, and not dependent on the view scale. If the view frame is larger than the view, white space will fill in around the view in the frame.

Display

The display is set to refresh only when the layout document is active, or if on the Always setting. Setting the display to When Active prevents the layout from refreshing when it is left open. This way, you can go to another document such as the view and perform work without the display being refreshed when you don't want it to be. This is a time saver when working with complicated layouts and detailed views.

Quality

The quality option works within the layout as the display option works outside the layout. If the quality option is set to Presentation, the view frame will always refresh as you zoom in, zoom out, or pan around the layout. If the quality option is set to Draft, the view frame will appear as a gray shaded box as you navigate around the layout, saving screen refresh time with complicated layouts.

Legend Frame

The legend frame tool works in a similar fashion to the view frame tool, in that it actively reflects changes in the Table of Contents in the view document. Whatever themes are visible in the TOC are placed by the legend frame tool in the layout, along with any symbolization or text labeling done in the view's legend editor. This "active link" is always on, once a legend frame is placed on the page.

The first step once the legend frame box is drawn is to select the proper view on which to base the legend, which should be the same view as in the view frame. The Legend Frame Properties dialog, shown in the following illustration, has two of the same options as the view frame: display and quality. The Display feature controls the refresh time of the legend when working in other ArcView windows, whereas the Quality option manages the on or off behavior of the legend within the layout.

Legend Frame Properties dialog box.

Changing Legend Size

Legend frames can be resized by selecting the graphic handles (the black squares) with the Pointer tool and dragging diagonally in or out from the center. A better way is to use the Symbol Window's Font palette, and select a type size and type style that match the remainder of your layout. The entire legend frame (symbols and text) will change size as you select larger or smaller type sizes.

Simplifying a Legend

The legend frame tool produces very basic one-column legends, which you will frequently want to change or enhance. One way of doing this is to "simplify" the legend from the Graphics menu, which breaks the single legend graphic block into its component pieces of symbols and text. Now the color and size of the legend symbols can be altered in the layout with the use of Symbol Window palettes. Legend pieces can be stacked into multiple columns in order to spread horizontally across a page, and text can be edited.

The use of Simplify should be the last step in the process of map composition in the layout document, because the simplify feature permanently breaks the active link between the legend frame and the view's Table of Contents. For more information on enhancing legends, see "Using the Custom Legend Extension" in Chapter 9.

Scale Bar Frame

The scale bar frame tool can be tricky to use, and requires some tweaking to produce the output you want. Much of its behavior is controlled by settings chosen in the View Properties dialog in the view document. This is shown in the following illustration. Map units must be set here for the scale frame tool to work at all, and if distance units are set, the same units will appear in the scale bar in the layout. If you chose kilometers for distance units in the view, the scale bar will default to kilometers in the layout.

Scale Bar Properties dialog box.

More so than other frame tools, the scale bar frame is sensitive to the size of the box drawn on the page. Overall, the scale bar should be small and unobtrusive as compared to the view and legend frames. Avoid the tendency to stretch the scale halfway across the width of the page. Instead, use a space two to three inches wide and not very tall, below the legend for the scale bar. After placing the box, make sure the scale bar is based on the correct view document and select a style for the map element.

Tweak the numbers for intervals and left divisions until you have a readable scale. You can reaccess the scale frame dialog at any time by double clicking on the scale bar, or by resizing the bar using the Pointer tool or the Font palette in the Symbol window.

North Arrow Manager

The North Arrow Manager, shown in the following illustration, offers eight standard North arrows and compasses that can be placed on the page. The North arrow, being a minor graphic element, should not take up much room on the page, and is often associated with and placed next to the scale bar.

North Arrow Manager.

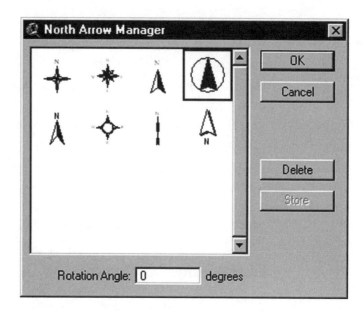

✓ **TIP:** *ArcView gives you the option of creating your own custom compasses or North arrows, from either the graphic tools in the layout or by importing a graphic via the picture frame tool. Either way is great to combine a directional compass with your company's logo for a nice professional touch to a map. To save the new graphic, use the Store North Arrows feature in the Layout menu. Individual graphics or text making up the North arrow must be grouped as one graphic prior to storing. Once stored, the new north graphic appears in the North arrow dialog box, and can be used or deleted from the North arrow list.*

The new stored North arrows are saved in an object database file (an ODB file) named *north.def*, which is placed in your default working directory, commonly *c:\temp*. If you want to easily access this file in the future, set your working directory to some other location (the ArcView installation directory *c:\esri* on Windows PCs, for example) before creating the North arrow. The working directory is set in the File menu in the view document.

One problem with North arrows is that they may not point to true North on some small-scale maps, because the North arrows are simple graphics, and are not live linked in any way to the map features in the view document. For example, you begin working with a theme in a view that contains the continental United States. After zooming into the Pacific Northwest, you create a layout using the Layout feature in the View menu. The resulting North arrow on the map will point to the top of the page, but true north will be 10 or 20 degrees to the left. To correct this, go into the North Arrow Manager dialog and set a positive rotation angle of several degrees, to nudge the North arrow graphic to the right, and toward true north.

Chart Frame

Charts can be placed on the map using the chart frame tool. Charts placed on the page will appear grayed out, unless the Chart document itself is open, along with the layout. Chart frames are live linked to the Chart document, so that changes to one will appear in the other. The Chart Frame Properties dialog box is shown in the following illustration.

Chart Frame Properties dialog box.

ArcView's charting function tends to be quite limited, so a better manner of constructing a chart for your map may be to create a chart in another program, such as Microsoft Excel, save it as a graphic, and import it into the ArcView layout using the picture frame tool.

Table Frame

The Table Frame tool allows you to place ArcView table documents on a page, but it has many limitations. Tables that have fields that total more than 80 characters wide will be truncated, and only the leftmost under the 80-character limit will appear. You can work around this limitation by going into the Table Properties dialog box (accessed from the Table menu when you have an ArcView table open) and making only the most crucial fields visible, while hiding the rest. The Table Frame Properties dialog box is shown in the following illustration.

Table Frame Properties dialog box.

Still, placing tables in the layout has its drawbacks. Redraw of the table frame content when you zoom in or out can be lengthy, unless the table frame properties are set to redraw only when the frame is active (selected with the Pointer tool). Printing time can be increased considerably as well by the placement of a table on the map. If you plan on needing many tables, charts, and graphs for your map documents, it may be worthwhile to explore the Seagate Crystal Reports software, which ships with ArcView. Seagate Crystal Reports has many tools that make producing graphic-rich reports much easier than those found in an ArcView layout.

Picture Frame

More than a dozen file formats can be placed on the layout page using the picture frame tool. This allows users to place graphics such as logos, aerial photos, digital camera pictures, and images from ERDAS Imagine software in the layout. The Picture Frame Properties dialog box is shown in the following illustration. Table 8-4, which follows, presents file types used in the picture frame.

*Picture Frame
Properties dialog
box.*

Table 8-4: File Types Used in the Picture Frame

Name	Extension	Type	Notes
Band Interleaved by Line	.bil	Raster	—
Band Interleaved by Pixel	.bip	Raster	—
Band Sequential	.bsq	Raster	—
CompuServe GIF	.gif	Raster	—
Encapsulated PostScript	.eps	Vector	Only PostScript Level I
ERDAS GIS	.gis	Raster	—
ERDAS LAN	.lan	Raster	—
IMPELL Bitmap	.rlc	Raster	—
MacPaint	.mcp	Raster	From the Macintosh OS
Nexpert Object Image	.nbi	Raster	From the Next OS
PostScript	.ps	Vector	Only PostScript Level I
X-Bitmap	.xbm	Raster	UNIX file format
Sun Raster	.rs	Raster	UNIX file format
TIFF Bitmap	.tif	Raster	Most common graphic format
Windows Bitmap	.bmp	Raster	Windows file format
Placeable Windows Metafile	.wmf	Vector/raster	Windows file format
Windows Metafile	.wmf	Vector/raster	Windows file format

Although the most common use of the picture frame tool is probably for placement of a logo, you can think more creatively here. For example, a series of images taken with a digital camera can be arranged around the map, with graphic lines pointing to their actual location. An aerial photo of a site can be placed beside a vector map of same area. Alternatively, a GIF image of a web page can placed on the map to show where more geographic information can be obtained from the Web. The picture frame tool's dialog box offers the Display and Quality features found in other frame tool dialogs.

Using Graphics Tools to Organize Elements

Although, strictly speaking, ArcView is not a graphic design program, the layout window contains several tools that greatly aid in organizing graphic elements on the page. These tools include the Size and Position feature, the Align feature, the Grouping and Ungrouping tools, and the neatline drawing tool. These tools come into play after you have placed most of the map elements in the layout with the help of the frame tools, and are ready to tweak the final appearance of the map.

Graphic Size and Position

The Size and Position feature in the Graphics menu gives you precise control over the location of all elements on the page. Any frame or placed graphic can be sized by typing in inches or metric equivalents for height and width. Likewise, the frame can be positioned a specific distance from the left or right, or top or bottom, edge of the page. The Size and Position feature is shown in the following illustration.

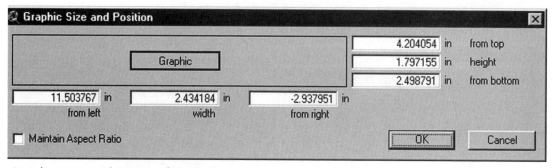

Graphic Size and Position feature.

Use this feature to set enough distance on one side of a map for binding a map in a report, organizing map elements to fit in a specific window on a page, or refining the layout of a map you need to use in a template.

Another set of features, Bring to Front and Send to Back, controls the position of the map elements in the layering or drawing order of the page. Selected elements sent to the back will draw first, and appear "underneath" elements brought to the front of the drawing order. Use these features to place filled boxes "behind" legend text, or to ensure that neatlines are brought to the front to overprint view frames.

Aligning Elements

When map elements are placed with the frame tools, they can be selected with the Pointer tool and generally moved from place to place. However, it can be difficult to line up graphics so that they appear organized visually, and the Snap to Grid feature in the Layout properties can prove insufficient for this purpose.

An easier method of organizing map elements is to use the Align feature in the Graphics menu. First select the graphics or frame you want to arrange, and then go to the Align feature. The Align dialog box gives you a series of vertical and horizontal controls, as shown in the following illustration.

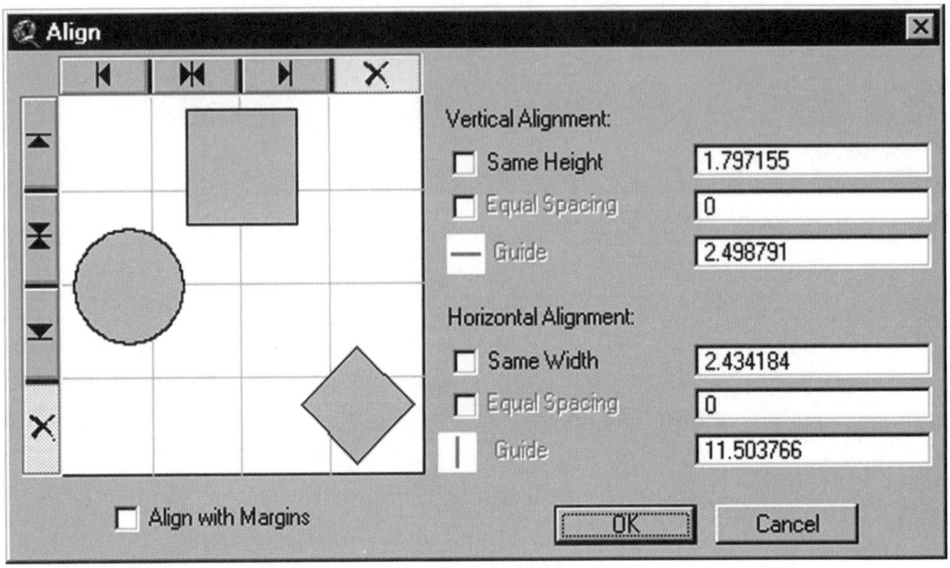

Align controls.

The Align feature's horizontal controls allow you to right- or left-justify or center the selected graphics as they relate to a guideline. The guide line can be automatically chosen by using the align buttons (the farthest right- or left-selected graphic defines the guide) or by physically typing in the guide's location, in the page units chosen in the Page Setup dialog. Page units are viewed as existing along an x and y axis with the 0,0 point starting on the lower left corner of the page. Align's vertical controls work in the same manner as the horizontal controls and guides.

There are two more align options available: the Same Width-Same Height feature and the Equal Spacing option. The Same Width-Same Height features need to be used with caution, because they will stretch your selected graphics until all graphics are the same width or the same height. These tools are best applied to graphic boxes, such as those in a simplified legend, and not to any element containing text. The Equal Spacing option is handy for vertically spacing portions of a legend or pieces of text, because it only alters the white space between map elements and not the elements themselves.

Grouping, Ungrouping, and Simplifying

The real power of the Align controls comes into play when used in conjunction with the Group, Ungroup, and Simplify tools. Grouping joins a series of graphic elements, so that they can moved as one object around the page, or sent to the back or brought to the front of the drawing order. Grouped objects can always be ungrouped.

Simplifying breaks up compound objects (such as legend, North arrow, or scale frames) into their constituent graphic pieces, allowing the user to change colors of a legend. Unlike grouping, simplifying cannot be undone, and the live link between a legend and a view is completely lost. However, once a legend is simplified and altered, the basic graphics can be aligned (left justified with equal vertical spacing) and finally grouped, to preserve the changes.

Neatlines

The Neatline button offers an automated means of placing graphic boxes around individual map elements, or around the entire page. Neatlines are useful as organizing tools (such as drawing a box around a legend), or as positioning tools (such as outlining the margin of a page or providing a cut line for a printed map on a plotter).

As with the Align tools, you first need to select, with the Pointer tool, the map elements you want to apply the neatlines to, and then click on the Neatline button. The Neatline Settings dialog, shown in the following illustration, allows you to automatically group the neatline with the selected object, precisely offset the neatline from the object, and set the line width and fill of the neatline box. Neatlines come in five single and multi-line styles and can have rounded corners or drop shadows.

Neatline settings.

It is good to freely experiment with the use of neatlines on the page, with the caution of avoiding the overuse of the feature, and ending up with a boxy, over-outlined map.

Drawing Graphics

The ArcView layout documents offers the basic graphic drawing tools for points, lines, and polygons, which can be placed anywhere on the page. Point symbol size, line weight, and fill color are all controlled through the Symbol Window palettes.

Extra graphics can add a custom touch to a map in several ways. Use thin, straight lines pointing to significant map features as "callouts," with text blocks explaining the importance of the feature, or draw circles or rectangles around areas of the map to be shown in more detail in another (presumably zoomed-in) view frame.

Text Tool

The Text tool is the primary means of adding a title to the map, and placing supplemental text such as the source statement. If you have created the layout automatically from the view document using the Layout option in the View menu, the name of your view window becomes the name of your map by default. Otherwise, you will need to add a title using the Text tool.

The Text tool in the layout document behaves in the same manner as the Text tool in the view, and is fine for placing short phrases of text for titles or margin information. It is not useful for naming map features, because if the view is changed, text placed over the view frame in the layout will not move with it.

The Text tool is also less helpful when trying to add larger blocks of text, such as source statements or descriptive text for an info graphic. Because the text must be typed into the small window within the Text Properties dialog box, shown in the following illustration, you have no way of knowing when to add a return at the end of a line to get the text to wrap.

Text Properties dialog box.

> ✓ **TIP**: *A faster way of adding larger blocks of text to your map is to write the paragraphs in a word processor and use the keyboard to copy and paste the text into the Text Properties dialog box. Text copied in this manner preserves the hard returns at the end of the lines as you want them to appear.*

Using the Template Manager

The ArcView layout provides the Template Manager, shown in the following illustration, as a handy means of storing and retrieving good map designs. This is particularly helpful when you are producing a map series in which all maps need the same look and feel of design.

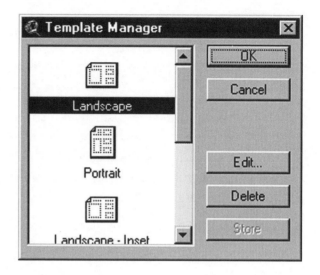

Template Manager.

You can start with a new, empty layout document, set the page size and orientation, and begin to place map elements on the page. To first place the view frame, in the View Frame dialog box, select Empty View. The legend and scale bar frames should all be based on this empty view. A partial title can be added (such as Map 1 or Figure 1 for the start of a series of graphics in a report), as well as a North arrow and a company logo.

When you have arranged the map elements to your satisfaction, use the Store as Template option in the Layout menu to save the template, making sure to use a unique name for the template file. This file can be accessed each time you construct a layout document automatically from the view window, using the Layout option in the View menu.

As with storing custom North arrows, the template file is saved as an object database file in your default working directory. It is best to set the directory to something like the ArcView installation directory before creating the custom template, so that the *template.def* file can be easily found for future use.

Summary

The ArcView layout document provides you with a host of powerful tools for placing your map product on a page, and arranging the map elements in an easy-to-read manner. The powerful tools will not supply you with a sense of map design, which needs to be developed from studying existing maps, knowledge of design rules, and from practice. Chapter 9 introduces you to extra tools and techniques for polishing your map design skills.

Chapter 9

Advanced Views and Layouts

Chapter 8 introduced the basics of design as applied to the map in the ArcView layout. This chapter addresses advanced techniques and tools applied in both the view and layout documents.

The techniques go beyond simple data symbolization into graphic design, with the goal of making your map more readable and visually attractive, setting it apart from other maps in our cluttered information world. The tools consist of a dozen free or low-cost extensions, all available on the World Wide Web. These extensions add increased functionality to the basic ArcView tools.

As with any advanced tool or technique, some of these take practice to learn their eccentricities, and they might not save you production time the first time you try to apply them. However, they do show how ArcView can be enhanced to produce high-quality map products.

Advanced Techniques in the View

GIS is primarily a technical discipline that has evolved out of database and computer-aided design roots. It can be disconcerting for many technical GIS specialists, while making a map, to confront the "art" in *cartography*.

Most of the techniques described in this section have little to do with data analysis, but can greatly improve the visual design of your map product. In fact, several of the techniques, such as generalization, can actually wreck your data, and should only be applied to real copies of a theme (remember to use Convert to Shapefile). Table 9-1, which follows, lists and describes extensions used with the ArcView view document.

Table 9-1: Extensions for Use in the ArcView View

Name in ArcView	File Name	Name in Extension Dialog	Producer
Geoprocessing Wizard	geoproc.avx	Geoprocessing	ESRI
Xtools	xtools.avx	Xtools Extension	Oregon Department of Forestry (Mike Delaune)
Generalize	generalize.avx	Generalize	ESRI
Legend-Smooth-Continuous	Legend-Smooth-Continuous.avx	COS.Legend-Smooth-Continuous	Robert Chasen
ColorRamp	ClrRamp0.avx	Color ramp	Bill Huber

Using Multiple Copies of the Same Theme

One of the more powerful techniques to master in the view is that of using the same theme twice, to sandwich other themes between backgrounds, outlines, or transparent hatch fills. Use the copy-and-paste feature in the Edit menu to make a copy of the theme in the view, and then symbolize the copy differently with the Legend Editor.

As an example of its most simple use, you might create a sandwich using a political jurisdiction theme twice. For example, use counties as a background, shown as a light colored fill with no outline, add a new theme on top of it, and then combine a copy of the counties theme on top with no fill and a dark black outline. This is shown in the following illustration.

Using multiple copies of a theme to sandwich another theme.

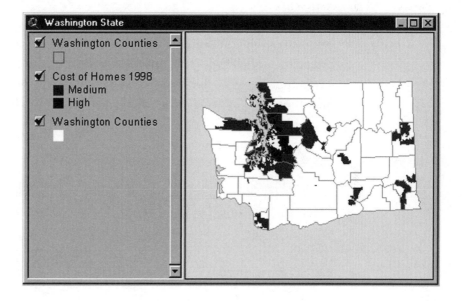

This technique is useful in many other situations, such as using transparent hatch polygons with no outlines as the center layer in a political jurisdiction sandwich to show data that crosses political boundaries. When using hatch fills, make sure to load the *carto.avp* PostScript fills to avoid printing problems.

Plate 8 in the color insert shows an example of the sandwich technique, as well as a use of buffers. The background theme is a county theme, followed by a darker colored buffer of the county boundaries. The buffered themes are sandwiched with yet another county theme showing polygon boundaries as thin, white lines with no fills.

Using Buffers for Cartographic Effects

The Create Buffers feature, introduced in ArcView 3.2, does a great job of buffering map features. Similar buffering tools are available in the popular Xtools extension, created by Mike Delaune at the Oregon Department of Forestry, found on the web at *http://www.odf. state.or.us/sfgis/*.

Buffering polygon features gives you a powerful means of showing boundaries of all types as a series of graduated color fills, rather than a confusing tangle of thick, black lines. Buffering effects can be divided into those useful for hydrographic features, and those applicable to boundaries on land.

Water Buffers

Use the Create Buffers feature to give the impression of deepening water surrounding land polygons. This is useful anytime you have maps with large bodies of water with islands, bays, or peninsulas, and you want to add a professional cartographic touch to the finished map, as shown in the following illustration.

The buffering process.

Before starting a buffer process, always check your View Properties, to ensure that map and distance units are set. Next, select Create Buffers in the Theme menu. When the Create Buffers dialog appears, select the theme with the land masses you want to buffer and click on Next. On the second panel, check the radio button for multiple rings, with three to five rings, and a spacing of 1 to 5 kilometers.

You will have to experiment several times with the spacing numbers to achieve the effect you want, depending on the map's scale. Click on Next to go to the final panel. Here, respond with Yes to dissolve barriers between buffered features, and check "Create buffers only outside of polygons" for land features. If using lakes, you would check "Buffer only inside polygons." It is best to save these buffers as new themes on your hard drive. Therefore, name the new theme and click on Finish.

When the new theme appears in the view's Table of Contents, double click on it to bring up the Legend Editor, which may already be set as a graduated color legend type. Click on the first symbol in the legend to bring up the Symbol Window. At this point, go to the Palette Manager, and load the *artist.avp* color palette from your ArcView symbols directory. *Artist.avp* will greatly aid in choosing colors for the water buffers. Now try different color ramps, starting from light blue and proceeding to dark blue to give the illusion of water depth. Water buffers are shown in the following illustration.

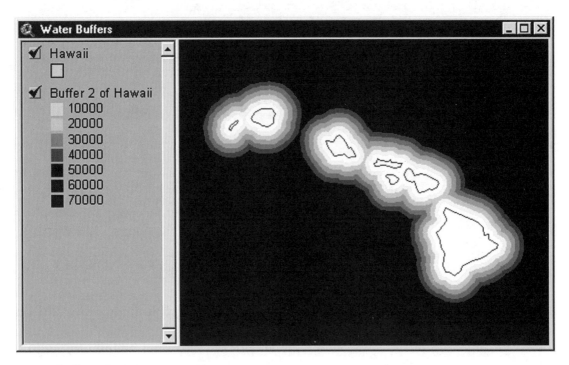

Water buffers shown in grayscale.

✓ **TIP**: *In general, use less fewer six buffers, and keep them close to the land boundaries for better effect. See plate 7 in the color insert for a color example of hydrographic buffers.*

Land Buffers

Political boundaries (such as states, provinces, counties, and census tracts) can result in a rat's nest of thick, dark lines on a map. Traditionally, cartographers have tried to solve this problem by using dashed and dotted lines, a technique that presents its own set of problems. In ArcView, dashed lines can be difficult or impossible to create correctly for polygon features.

Buffering political boundaries is a great way of avoiding the use of dashed lines for political jurisdictions, while minimizing line work and clutter on the map. You go about making them in the same manner you would to produce hydrographic buffers, with three differences.

For rings, select one ring only, about a half mile from the political boundary you want to buffer. Then select "Only inside the polygon only" for where to create the buffer. Make sure to save the buffer as a new theme on your hard drive. The resulting buffer is a single polygon, which in the Legend Editor should be symbolized with a fill but no outline. Land buffers are shown in the following illustration.

Land buffers shown in grayscale.

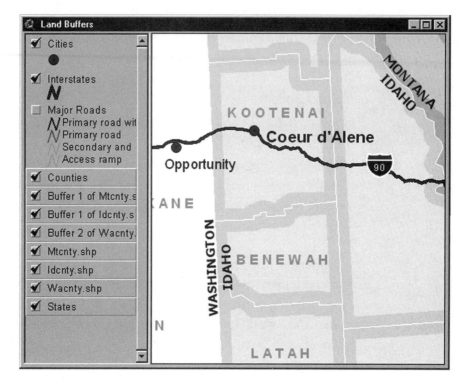

✓ **TIP**: *For the fill of the buffer polygon, choose a color slightly darker or lighter than the background fill of the political jurisdiction you have buffered. Use either the* artist.avp *palette for color selection, or keep the hue and value the same in the Custom Color feature of the Fill palette, and change the saturation either up or down.*

As a final touch, try using the sandwich technique and place another copy of the political jurisdiction on top of the buffer polygon, and symbolize this copy with no fill and a thin white line. For an example of this technique, see Plate 8 of the color insert.

Aside from boundaries and depth effects, the buffer technique can be used to call attention to selected features in line work such as rivers, roads, and utility corridors. Try experimenting with buffers and colors to create "glowing line" effects at times you need to make a line feature more noticeable.

Generalizing and Dissolving Features

Frequently, detailed GIS data themes are overkill for simply showing the location of features on a map. Even with 1:100,000 data (which is moderate to small scale), a polygon or line theme for such data as counties or rivers can appear jagged on a page-size layout. At other times, themes with thousands of vertices can fail to print on printers with limited memory. In the layout, the page may fail to export as EPS or JPEG, or if it does, the export may be unmanageable in a graphics program.

Some of these problems are solving themselves as desktop PCs are sold with more powerful CPUs and hundreds of megabytes of RAM on board, allowing large data files to be more easily manipulated. However, there are two means within ArcView of minimizing problems of large files. The first is to use the Geoprocessor or Xtools extension to dissolve unnecessary attributes in a particularly complicated theme. The second is to use the Generalize extension from ESRI.

Dissolving on a theme's attribute can be thought of as a means of simplifying both the attributes and the graphics of a theme. If the graphics are simpler, the attributes may be easier to present on a map, as well as easier to print. To perform a dissolve, load the Geoprocessing extensions from the Extensions dialog, and select Geoprocessing Wizard from the View menu.

You will be prompted to choose a geoprocessing operation (dissolve), and then asked which theme to perform it on. After picking an attribute within that theme to be dissolved, save the theme as a new shapefile on the hard drive. Last, you will be asked if you want to calculate statistics in the dissolve operation. After clicking on Finish, the new dissolved theme is added to the top of the Table of Contents. Remember, you are throwing detail out to get a more simplified graphic. The geoprocessing dialog for dissolves is shown in the following illustration.

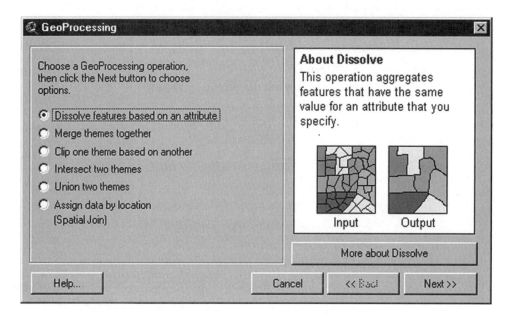

GeoProcessing dialog for dissolves.

There are numerous situations in which you might consider using a dissolve. Frequently, with political jurisdictions, you may not need the detail within a census track or a county, and you might not have access to a theme of the next larger political unit. Examine the attribute table of the theme in question, and often you can dissolve on the field that has the next higher political unit, resulting in county or state outlines, and a simpler map. Another good example concerns land-cover or vegetation maps.

A vegetation map may have a dozen classifications for coniferous forest that you are not interested in. All you want is to show the distribution of coniferous forest. Add a field to the theme's attribute table named Forest Type, and use the Field Calculator to label all subtypes as Coniferous Forest. Then, go to the Geoprocessing Wizard and dissolve on Forest Type to produce a much more simplified theme. An example of this is shown in the following illustration.

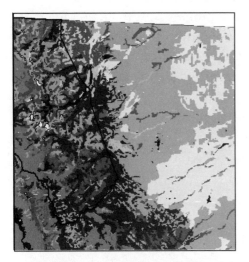

Land-cover map before dissolve.

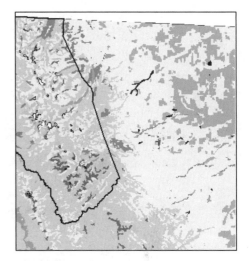

Land-cover map after dissolve.

Results of a dissolve process.

Generalizing a theme is the other means of simplifying a theme. Rather than removing attribute features, generalizing removes actual vertices or control points within a feature, using the Douglas-Peucker algorithm. In addition, the generalization process does not produce a new theme, but irrevocably alters the data in your original theme.

✗ **WARNING**: *Be sure to make a copy of the data using Convert to Shapefile before trying this one!*

Load the Generalize extension from the Extension dialog. The Generalize button will appear on the far right of the view's button bar. Before you can apply the button to the selected theme, however, you need to select the features you want to generalize (using the Select Features tool in the View) and Start Editing. Then, click on the Generalize button, and the Generalize dialog appears, which is shown in the following illustration. The dialog prompts you to enter a tolerance in distance units (set in the View Properties window). Enter the tolerance and click on Apply. The dialog box remains until you close it, showing the number of vertices before and after generalization.

Generalize dialog box.

Generalize is a particularly convenient function when you want to simplify something such as a highly detailed hydrographic theme that would otherwise have lines depicting every twist and turn of a river or stream. This process would allow you to show a more generalized path (location) of the river, an example of which is shown in the following illustration.

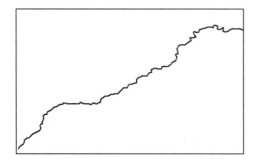

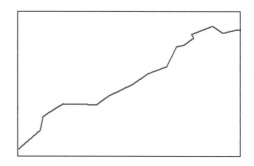

River before generalization.

River after generalization.

Results of using Generalize on a river line theme.

> ✓ **TIP**: *Generalize highly detailed hyrdographic themes you might use commonly in small-scale maps, and then store them along with other base data in views saved with the ArcView ODB extension.*

Legend Tools

There are quite a few Avenue scripts and extensions that aid in making legends. The Legend-Smooth-Continuous and ColorRamp extensions are two of the more interesting ones. These extensions are discussed in the sections that follow.

Smooth Graduated Color Legends

When confronted with dozens of classes in a graduated color legend, ArcView may repeat some colors or have big color jumps between classes. This can give the legend a choppy look. The Legend-Smooth-Continuous extension solves this problem by adding a "Smooth Gradations" feature to the Theme menu in the View window. This feature smooths out the color classifications in any graduated color polygon theme or GRID raster theme.

Load the extension from the Extension dialog, make your graduated color theme active, and select Smooth Gradations from the Theme menu. After doing this process in the view, go to your layout document, and the extension will have added a button to the right of the frame tools on the toolbar. Use this tool to draw a small rectangle on the page, and the extension will produce a continuous legend bar for your graduated color theme, rather than a series of disconnected boxes. An example of this is shown in the following illustration.

Annual Precipitation

1 inch   250 inches

A smooth, continuous legend bar shown in grayscale.

Legend-Smooth-Continuous is great for themes such as precipitation or elevation data. See Plate 10 in the color insert for an example of using this tool for a legend bar.

Custom Color Ramps

ArcView's ColorRamp extension gives you several options for creating custom color legends. ColorRamp, once loaded, adds a ramp button to the right of the De-select button on the view's button bar. Make a theme active in the Table of Contents, and click on the color ramp button. An entirely new legend editor appears, featuring its own color picker.

Color spaces can be switched from ArcView's native HSV color space to the more common RGB or LAB color spaces. Color ramps appear as a vertical band on the left of the ColorRamp dialog box, and custom colors can be set for either end of the ramp, or at any intermediate point. Final ramps can be saved and quickly loaded for use with other themes. The ColorRamp controls are shown in the following illustration.

ColorRamp controls.

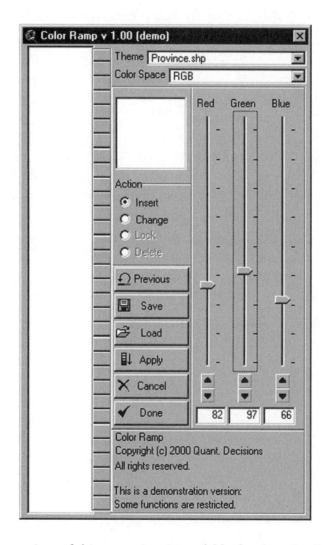

A free version of this extension is available for downloading from the ESRI ArcScripts page. A full-fledged version is sold on the Quantitative Decisions web site at *http://www.quantdec.com.*

Advanced Techniques with Extensions in the Layout

If you go the ArcScripts page on the World Wide Web at *http://www. esri.com*, you will find hundreds of free extensions and scripts that add functionality to nearly every aspect of ArcView. The sections that follow discuss half a dozen of the better ones, which can really enhance working in the layout.

One note on adding many extensions to an ArcView project: Most freeware extensions work as advertised when one or two of them are added to a project. How they work when you have a dozen different extensions loaded at once can be another matter. All of the extensions described here have been tested by the author and work very well by themselves. Therefore, do not be afraid to download a new extension and experiment a little. You will find that these tools really help in many of your common mapmaking tasks.

Custom Legend Tool

Legends are the second most important element in your map composition, yet with basic ArcView it is difficult to arrange them properly to show what you want. The standard legend frame tool in the layout produces only single-column legends containing every visible theme. For anything fancier, you need to simplify the legend and move graphics and text, which ends up breaking the live link with the view.

The Custom Legend Tool extension, called Legend Tool in the Extensions dialog and *legends.avx* in the *ext32* directory, is produced by ESRI and comes with ArcView 3.2. The extension enables you to produce quick, multi-column legends. Unlike the legend frame, the custom legend tool does not produce live-linked legends, but only graphic objects that will not change as your view document is updated.

For this reason, it is smart to wait toward the end of your map production process in the layout before putting a lot of effort into constructing the perfect legend. The Custom Legend tool gives you a lot of flexibility in the size and shape of your legend, so it is good to have an idea of where you want to place your legend on the page. In addition, spend some time experimenting with the tool's options to see how the various results will fit.

After loading the extension, which appears as Legend Tool in the Extensions dialog, the tool appears as a button to the right of the frame tools on the toolbar GUI. Clicking on it brings up the Custom Legend Wizard, the first panel of which contains introductory text describing the legend-making process. The rest of the wizard leads the user through a five-step process of designing the legend. The steps in this process are discussed in the sections that follow.

Step 1: Establishing Legend Parameters

Here you select the view window on which to base your legend. Visible themes appear in the left-hand menu. You must move themes from the left- to the right-hand column to include them in the legend. Remember from the legend discussion in Chapter 8 that not all themes, especially obvious base layers, need to be included in a legend, and this panel gives you the opportunity to pick and choose the relevant ones. The illustration that follows shows steps 1 and 2 of the process.

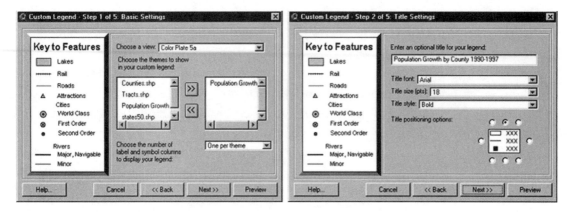

Steps 1 and 2 of the Custom Legend tool process.

Also on this panel is the option for number of columns in the legend, with the default being one column per theme. This is a tricky feature, which does not always work the way you think it might. First, the tool will grab all of the legend features for one theme, even though they might be hidden using Hide/Show Legend under the Theme menu in the view.

Next, if the themes you select have longer individual legends, the Custom Legend tool may not keep all of them in the same column, and end up splitting them with an adjacent column. Make sure to keep using the Preview option in the lower right corner of the wizard panel to check on what the new legend will look like. It may be impossible to split themes evenly among certain numbers of columns, and you will have to simplify the resulting graphic and split up legend features manually.

Step 2: Providing a Title, and Establishing Type Style and Size

On the next panel, you may type in an optional title, and position it in relation to the rest of the legend. More importantly this panel allows you to pick a type style and size for the entire legend. For some peculiar reason, the type size always defaults to a large number such as 18 pt. This is much too large for anything under a poster-size map, and should be reset to 10 or even 8 pt for a letter-size composition.

Step 3: Establishing Parameters of the Legend Box

The third panel controls the legend box. If you do not want a legend, you will have to simplify the resulting graphic and select it in order to delete it. Three border styles are offered for the legend box (less than the number available under neatlines), as well as line weights and colors, polygon fill colors, drop shadows, and rounded corners. Once again, the line weight defaults to 0.1 pt, a nearly invisible width on many printers. Make sure to bump it up to at least 0.5 pt. Be careful not to be carried away with the drop-shadow feature, overuse of which can detract from your map. The following illustration shows steps 3 and 4 of the Custom Legend tool process.

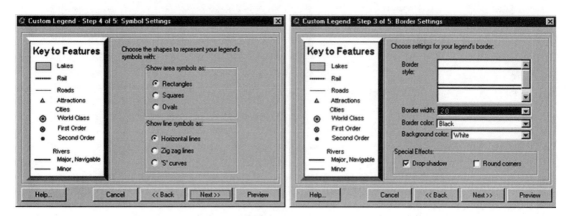

Steps 3 and 4 of the Custom Legend tool process.

Step 4: Establishing Polygon and Line Representation

Panel four gives you more options for how the symbols (polygons and lines) will be represented in the final legend. Polygons can be represented as rectangles, squares, and ovals, and lines as straight horizontal objects, zig-zags, or S curves. Strangely, these are somewhat different settings than those in the TOC Style dialog in the View menu.

Step 5: Establishing Spacing Parameters

Last, step 5 controls the spacing between columns, and within legend features, such as theme names and symbols. Step 5 of the Custom Legend tool process is shown in the following illustration.

Step 5 of the Custom Legend tool process.

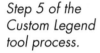

With all of these options, the Custom Legend tool can be time consuming the first few times you apply it, but it is well worth the effort to learn to use. You will produce better map legends. Table 9-2, which follows, lists and describes extensions used with the ArcView layout document.

Table 9-2: Extensions for Use in the ArcView Layout

Name in ArcView	File Name	Name in Extension Dialog	Producer
Custom Legend Tool	*legends.avx*	Legend Tool	ESRI
Graticules and Grids	*gratgrid.avx*	Graticules and Measured Grids	ESRI
Mapper	*zmapper.avx*	Mapper zmapper.odb	Howie Sternberg
Save and Restore	*odb.avx*	ArcView ODB Extension [v1.2]	ESRI
Overview	*overview.avx*	Overview Utility	ESRI
Nudger	*vfe.avx*	View Frame Extent Nudger	ESRI

Graticules and Measured Grids Extension

Everyone has seen maps in atlases with latitude and longitude lines draped over the continents. The Graticules and Grids extension offers this feature within the ArcView layout, with some limitations.

Graticules are latitude and longitude lines and labels that give you geographic locations in degrees, minutes, and seconds. A measured grid is a series of tic marks that measures linear distance on your map display.

⇢ **NOTE**: *The graticule portion of this extension only works with data stored in decimal degrees or "unprojected" (sometimes called a geographic projection). Any views with themes that are in a projection (such as Albers, UTM, state plane, and so on) will not work with the graticule feature. Projected data will work with the measured grids option, however. For more on projections and coordinate systems, see Chapter 7.*

The extension works in much the same way as the Custom Legend tool, providing a wizard-type interface that steps you through the process of creating graticules first, and grids second. The final result is a graphic that is not linked to your views, but can be ungrouped and further refined with the Symbol Window palettes. The process of using the Graticules and Grids feature is shown in the following illustration. The steps involved in this process are discussed in the sections that follow.

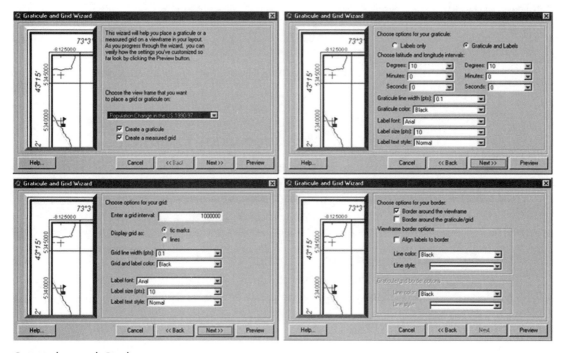

Graticules and Grids process.

Step 1: Establishing View Frame Parameters

The first panel prompts you to select a view frame on which to base the process, and gives check box options for creating a graticule, or a grid, or both. If you choose both, the graticules and grids will be grouped in the final graphic.

Step 2: Establishing Graticule Parameters

Next, set the options for the graticule. You can pick a variety of spacing options for the latitude and longitude lines, by degrees, minutes, and seconds. As with many ArcView graphic defaults, the line width is 0.1 pt and the label size is 10 pt. Make the line weight thicker and the label point size a touch smaller, so that the lines can be seen and labels do not overpower the map.

Step 3: Establishing Grid Parameters

The options for the grid follow step 2. A grid interval is set first, in meters, with its origin in the center of the view frame. The grid can be displayed as either tic marks (large plus signs) or lines. The grid line weight and color, and the grid label features, can be determined as well. If you are displaying both graticules and a grid, try using the grid as tic marks rather than lines, or choose a different line color for one to help the map reader distinguish between the two.

Step 4: Establishing Border Parameters

The final panel controls the border line settings, offering five border line styles and a variety of colors. Borders can be limited to the extent of the view frame, with labels placed outside the lines, or borders can be placed around the labels for the graticule and grid, enclosing the entire graphic. An example of graticules used on a small-scale map is shown in the following illustration.

Use of graticules on a small-scale map.

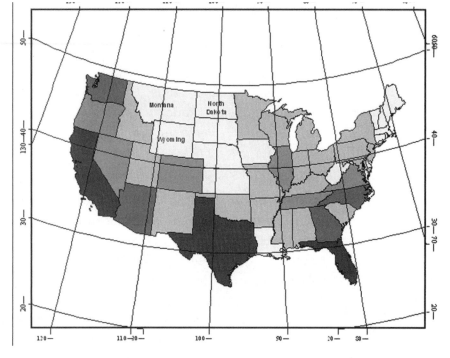

Latitude and longitude lines add a particularly nice touch to many small-scale maps, and a measured grid is very helpful for gauging distances on larger-scale maps. The Graticules and Measured Grids extension should be a valuable supplement to your ArcView toolbox.

Mapper Extension for Custom Scale Bars and Text

The Mapper extension adds several very neat features to your views and layouts. Mapper is a little tricky to install, because it comes with an ODB object that has to be located in the default user directory. Therefore, follow the directions in the *readme* text file carefully before trying to load Mapper in an ArcView project.

In the View window, the Print button permits you to create a quick and very simple map of the View window. Just click on the button, select a couple of options in the dialog boxes that follow, and the map is ready to be sent to the printer, complete with a simple scale bar, North arrow, title, and date.

In addition, two tools added to the Layout toolbar really make Mapper stand out. The first is a series of Text tools included as a drop-down tool list, just to the right of the frame tools. One text tool places text files as graphic blocks on the page without using the keyboard copy-and-paste trick (used with the Text Properties box, previously described).

The three other options of this tool place selected records from attribute tables on the page, with a variety of options. The first table text tool simply places the selected records in one live-linked table frame in the layout. The standard ArcView table frame tool places the entire visible table on the page, not just selected records.

The second text tool places selected records on the page and summarizes numeric fields for the selected set. The last text tool places selected records as vertical columns, rather than horizontal rows, in the layout. Aside from the enhanced text and table tools, Mapper provides a series of custom scale bars, based on common scales for maps. Table 9-3, which follows, lists Mapper scale bar numbers and their equivalent scales.

Table 9-3: Common Scale Bar Numbers/Scales in the Mapper Extension

Scale Bar Number	Scale	Scale Bar Number	Scale
12	1:12,000	100	1:100,000
24	1:24,000	125	1:125,000
35	1:35,000	150	1:150,000
50	1:50,000	200	1:200,000
75	1:75,000	250	1:250,000

The resulting scale bars are attractive, showing miles and kilometers, as well as a note on "miles to the inch" at that scale. The scale bars are static graphics and are not live linked to your view, so you should not resize them. You must also set your view frame to the same scale when placing your map display, to make sure the scale of the map matches the scale of the scale bar. However, this is still a very nice touch for a professional-looking map. The following illustration shows Mapper custom scale bars.

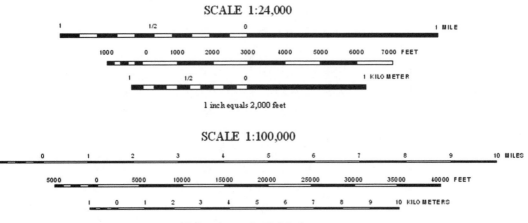

Custom scale bars from the Mapper extension.

Saving and Restoring Documents Using the ODB Extension v1.2

Things can get complicated when you are trying to produce a set of similar maps for a report. Often, many of the maps have the same map area, such as a single city or state, or perhaps a study area or sales region. The only feature that changes from map to map may be a single theme you are turning on or off in the view, or are symbolizing differently using the Legend Editor. Typically, you might base several map layouts on a single view window, and turn off the live link feature in the view frame properties.

Invariably, somewhere around the sixth or seventh layout, you are so confused you are not sure what to change next. In addition, because you turned off the live link, you cannot go back and easily edit the views for previous maps. This is where you need the ArcView ODB extension.

ODB stands for object database, and every document in ArcView (including views, tables, charts, layouts, and scripts) can be treated as individual objects. Once the ODB extension is loaded from the Extensions dialog, any ArcView document can be copied and pasted within the project, or stored as an individual ODB object outside the project. In an entirely different project, the object can be restored and worked on just like it was created there.

This is an incredibly handy feature for producing a map series. Start with a single view (U.S. Population, for example) and one layout live linked to it. Load the ODB extension, select the layout in the Project window, and select Copy from the Save and Restore menu. The extension will prompt you "Do you want to make new copies of the Layout's Views?" Click on Yes and you have a new view copy and layout copy, based on the originals.

To keep the documents straight, rename the originals "1 - U.S. Population View, Standard Deviation Legend" and "1 - U.S. Population Layout, Standard Deviation Legend" and the copies "2 - U.S. Population View, Graduated Color Legend" and "2 - U.S. Population Layout, Graduated Color Legend."

The numbers in front of the names allow the Project window to keep the copies in a numerical list with number 1 on top for easy access, as shown in the following illustration. Successive layouts can be copied, renamed, and tweaked to create a series of similar maps, all the while allowing you to go back to a previous view and layout.

A Project window organized by numbers for copied views and layouts.

Chances are, most of the maps you make are of the area in which you live and work. It can get old loading the same series of themes for base layers into a new project again and again. Using the ODB extension, make one or two view windows with most of your common base layers symbolized the way you like them, and store them as ODB objects on your hard drive. The next time someone requests a map of your area with their new data added, just load one of your base data ODB object views or layouts and the map is almost done.

View Frame Extent Nudger Extension

This extension produced by ESRI helps a lot with one of the irritating tasks in the ArcView layout: getting the view frame to show *exactly* what you want. Anyone who has produced many maps in Arc-View has had a project for which the view just does not fit on the map the way you want it to. You end up resizing the View window countless times, while trying different view frame properties (such as Clip to View) and different sizes, but still coming up short.

The Nudger is a button found on the layout's button bar. After selecting a view frame with the Pointer tool, click on the diamond icon button to bring up the Nudger dialog, shown in the following illustration. You select measurement units, and a nudge distance, and then use the convenient directional arrows to move the view within the view frame. This is a great way of fine tuning your map display within the layout, while avoiding the endless resizing-the-view routine.

View frame Nudger dialog.

Overview Extension

The Overview extension from ESRI loads into the View window, but its real usefulness shows up in the layout document. Called the Overview Utility in the Extensions dialog box, once loaded, this utility adds a feature to the View menu named "Create overview." When you create an overview, a simplified view window appears, without its Table of Contents.

The Overview window contains a rectangle (see following illustration) that moves as you zoom in and out of your main view, tracking the extent of that window. This is a handy feature if you have small-scale map data in your view and you want to know where you are as you navigate around it.

Overview window used as a locator.

It is in the layout document the Overview extension becomes really useful. Set up a page using the main View window, and the Overview window is available as a view frame for an instant locator map. This is particularly useful when you are producing a series of maps from small-scale data, and you need to keep your map audience informed of where each map in the series is located, in the larger scheme of things.

Summary

Once you have mastered the basics of views and layouts, it is time to branch out with new tools to increase your mapmaking creativity. The tools in Chapter 9 should help you produce map products that stand out from the crowd. Chapter 10 goes one step further, showing you how to begin using imagery and surfaces in your map composition.

Chapter 10

Using Raster Data in the View

0   Miles   500

Raster data, such as images and shaded relief surfaces, add a great deal of information to an ArcView map. So great that it is easy to overwhelm the map reader with information that may look interesting but has little to do with the map's purpose. Still, a well-composed map with a shaded relief background can be an eye-catching way of getting your point across.

This chapter looks at how basic ArcView handles raster data, and examines several techniques for visualizing raster data with two ArcView extensions, Spatial Analyst and 3D Analyst. Rather than being an exhaustive description of handling raster data, which can easily be the topic of its own book, this chapter will give you ideas and techniques on how to use aerial photos, scanned maps, imagery, and topographic surfaces in your map products. For a more extensive discussion of raster information, see *Raster Imagery in Geographic Information Systems* by Morain and López Baros (OnWord Press).

Types of Raster Data

All raster data stores information in the same basic manner, in cells or pixels arranged in rows and columns. The cells or pixels can have a range of values that correspond to many attributes of the environment, such as reflectance or absorption of light by the earth's surface, elevation above sea level, and other attributes derived from vector themes. The sections that follow describe the various categories of raster products.

Satellite Imagery and Aerial Photography

In recent years there has been a virtual explosion in the availability of imagery of numerous types that can be added to ArcView projects. The most common are digital aerial photography and satellite imagery. Several other, more specialized, types of remote sensing, such as radar imagery, are becoming more common, but are not as widely used. A Landsat satellite image is shown in the facing illustration.

Imagery of both types can either be captured in the visible light spectrum (as seen by the human eye) or recorded by multi-spectral scanners on board spacecraft or aircraft that go beyond the visible spectrum into the infrared or ultraviolet.

Much of the multi-spectral imagery must be purchased and then analyzed in an image processing program to some extent before it can be used for map production. Other images, such as digital orthoquads (DOQs), can be ordered and used "as is" as great backgrounds for ArcView maps, showing realistic large-scale detail on the ground.

A Landsat satellite image in ArcView

Scanned Images

Scanned images consisting of paper maps, blueprint construction drawings, or historic aerial photos can be of great value for mapping. ArcView users can scan images on paper into their computers themselves, and use several extensions to register the images to geographic coordinates. Alternatively, scanned map images are available commercially, such as DRGs, an example of which is shown in the following illustration.

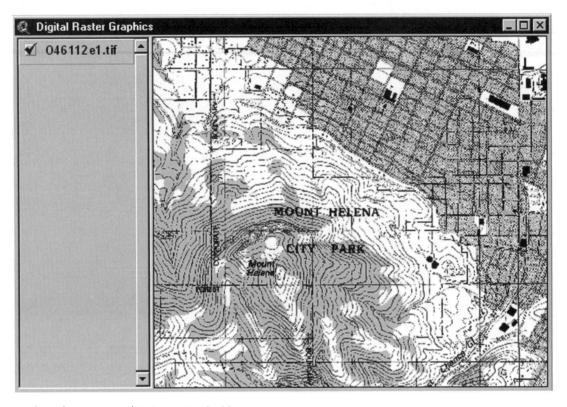

A digital raster graphic image in ArcView.

Digital raster graphics, called DRGs, are scanned images of USGS paper topographic maps that can add a wealth of background detail that many people instantly recognize as a "topo" map.

ArcInfo GRIDs

ArcInfo GRID files are really not images, but raster GIS files that can represent a wide variety of features. One of the most common GRID file types is the digital elevation model (DEM), the pixels of which reflect elevation above sea level. DEMs can be used to produce stunning shaded relief backgrounds for ArcView maps. Basic ArcView treats GRIDs as images, but ArcView Spatial Analyst can fully manipulate these raster GIS files. An example of a DEM is shown in the following illustration.

A digital elevation model added as an image file.

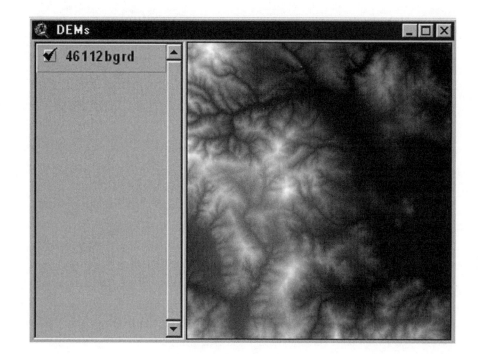

Image Data Types

Although raster data may seem simple to start with, it can get complicated very quickly when dealing with different forms of raster data. Raster data, including ArcInfo GRIDs, can all be treated as images, and images come in four basic types. The types are segregated by the amount of information each pixel can hold in bits. This is sometimes called *pixel depth* or *bit depth*. The more bit depth, the more information an image can contain.

Monochrome Images

Monochrome images are the simplest type of images. These are also called bitmapped or black-and-white images. In a monochrome image, there is only 1 bit of information per pixel. An individual pixel can either be black (a value of 1) or white (a value of 0). Monochrome images, like black-and-white computer monitors, are not used often, but can come into play when performing some image analysis.

Grayscale Images

Grayscale images have the same bit depth as indexed color images (8 bits per pixel) but use it to store 256 shades of gray. Values in a typical grayscale image range from 0 (black) to 255 (white).

Grayscale images are popular in mapping when used with simple shaded reliefs and many aerial photos. By minimizing the background image as a grayscale, the cartographer can use color in the foreground to emphasize different themes.

Pseudocolor Images

Pseudocolor images, or as they more commonly called, indexed color images, contain 8 bits of information per pixel. This makes 256 colors available in the image, which ArcView and ArcInfo store in a color-map or index file.

Indexed color images are useful for situations in which the full range of true color is not needed (for example, for scanned maps). An indexed color image will be considerably smaller in file size than a true color image, making it easier to manipulate and place on web pages.

True Color Images

True color images have 24 bits of information per pixel, allowing the image to hold 16.7 million colors. This enormous amount of color is very useful for portraying land-cover and vegetation distribution, but makes the imagery more difficult to manipulate, requiring faster computers and heftier printers. With the inevitable progression of Moore's law providing use with faster CPUs, and a host of new satellites in orbit, full color imagery is increasingly common in mapping.

True color images are also called multi-band images, in that the 24 bits of information are broken up into three 8-bit bands (channels). ArcView's Image Legend Editor allows you to manipulate each band individually. Some multi-spectral satellite images may contain as many as 16 bands. As a result, indexed color and grayscale images are called single-band images, and show up as such in the Image Legend Editor. Table 10-1, which follows, summarizes image data types used with ArcView.

Table 10-1: Image Data Types Used in ArcView

Name	Pixel Depth	Number of Bands	Notes
Monochrome or bitmapped images	1 bit per pixel	Single band	Black-and-white only. Not a grayscale image.
Grayscale images	8 bits per pixel	Single band	256 shades of gray. No colormap file.
Pseudocolor or indexed color images	8 bits per pixel	Single band	256 colors available, stored in a colormap file.
True Color images and multi-band images	24 bits per pixel	Multi-band	Each pixel has three 8-bit bands for storing information.

Basic ArcView Used with Raster Data

ArcView as it comes out of the box is primarily a vector GIS program, but it does give you several basic tools for manipulating images. You cannot permanently change image attributes or perform image analysis, but you can digitize features by drawing over the image, and you can include images as backgrounds in your maps.

You add images to a view with the Add Theme button. When the Add Theme dialog box appears, you must switch the toggle for Data Source Type from Feature Data to Image Data. Seven types of imagery are supported by basic ArcView, and an additional eight types can be loaded with the help of the free extensions that come with the program. Table 10-2, which follows, summarizes image formats used with ArcView.

Table 10-2: Image Formats Used with ArcView

Abbreviation	Format Name	Extensions Needed/Notes
ADRG	ARC Digitized Raster Graphics	ADRG Image Support Extension
BMP	Windows Bitmapped images	None needed.
BSQ, BIL	Band Interleaved Sequential, and BIP Line and Pixel images	None needed.
CADRG	Compressed ARC Digitized	CADRG Image Support Extension Raster Graphics
CIB	Controlled Image Base	CIB Image Support Extension
ERDAS	ERDAS .lan and .gis files	None needed.
GRID	ArcInfo GRID raster files	Can view GRIDs in basic ArcView. Can manipulate GRIDs with ArcView Spatial Analyst Extension.
IMAGINE	ERDAS Imagine .img files	Imagine Image Support Extension
IMPELL	IMPELL Bitmap .rlc files	None needed.

Abbreviation	Format Name	Extensions Needed/Notes
JPEG	Joint Photographic Experts	JPEG (JFIF) Image Support Extension Group *.jpg* files
MrSID	Multi-resolution Seamless	MrSID Image Support Extension Image Database *.sid* files
NITF	National Image Transport Format	NITF Image Support Extension
Sun raster files	Sun OS *.rs, .ras,* and *.sun* files	None needed.
TIFF	Tagged Information File Format *.tif, .tff,* and *.tiff* files	None needed.
TIFF 6.0	TIFF 6.0 files	TIFF 6.0 Image Support Extension
TIFF/LZW	LZW compressed TIFF files	Need optional software from ESRI.

Image Legend Editor

After an image is added to a view you can double click on its name to bring up the Image Legend Editor. This legend editor is considerably different from the standard legend editor, but it contains many tools common to image processing programs.

> ◆ **NOTE:** *In basic ArcView, the program treats ArcInfo GRID files as single-band images, and only allows you to manipulate them using the Image Legend Editor. With ArcView Spatial Analyst, GRID files can be loaded as GRIDS or images. When loaded as GRIDS, you can symbolize them with the standard ArcView Legend Editor as Graduated Symbol Legend types.*

When the Image Legend Editor opens, it immediately lists to the left the number of bands available in the image. For a multi-band image, you can manipulate all bands at the same time or treat each band as a single-band image. The Image Legend Editor is shown in the following illustration.

*Image Legend
Editor.*

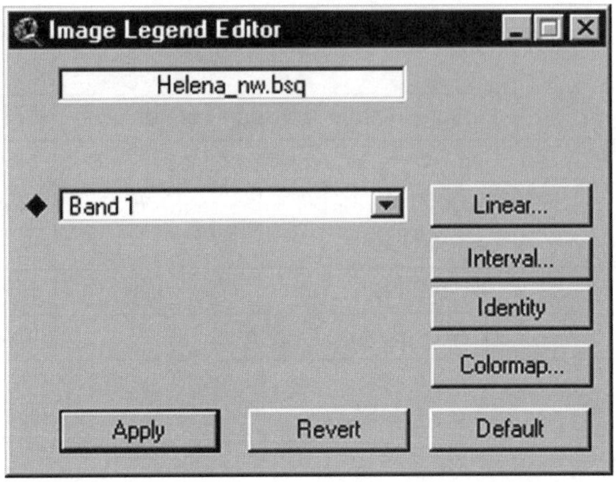

To the right in the editor are four simple image processing tool buttons: Linear Lookup, Interval, Identity, and Colormap. Across the bottom are the three application buttons: Apply, Revert, and Default. The three application buttons, respectively, apply the changes made with the processing tools to the image, revert to the last applied change, and revert to the default image as it first appeared when added to the view.

Linear Lookup Tool

The Linear Lookup tool, shown in the following illustration, controls the brightness and contrast of the image. The linear control panel appears as one graph for single-band images, and as three graphs for multi-band images, corresponding to the red, green, and blue colors. A straight diagonal line appears on the graph, showing input pixel values mapped on the x axis, and output pixel values on the y axis. If your image has a statistics file associated with it, a curved, bell-shaped line will be displayed in the background, depicting the mean of the pixel values at the top of the curve and dropping to 2 standard deviations on either side of the mean.

Linear Lookup tool.

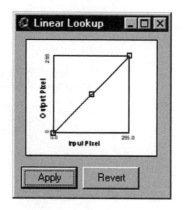

You can change the brightness and darkness of the image by moving the diagonal line back and forth using its center control point. Drag the line to the right to increase brightness, and to the left to make the image darker.

To increase the contrast between the lowest and highest pixel values, drag the endpoints of the line to make the graph steeper. To decrease contrast, drag the endpoints of the line to make the graph flatter. Click on Apply to show the changes, or select the Default button to go back to the original image.

✓ **TIP:** *Images can frequently be too dark to allow overlying themes or text to be readable. Use the Linear Lookup tool to lighten the image. Increasing contrast can have the same effect for much of the image, while causing darker (more saturated) features to stand out.*

Interval Lookup Tool

Less useful for many single-band images is the Interval Lookup tool, shown in the following illustration. The Interval Lookup tool provides a slider gauge, which will split the image into an equal interval classification with between 1 and 20 classes, once you select Apply.

Interval Lookup tool.

This tool works best with GRID files and multi-band images. The Interval tool can split the large number of elevation values in GRIDs, such as digital elevation models, into just a few classes with the Interval tool, giving the GRID image a contour effect. This is the best that can be done to show relief with raw GRIDs without the Spatial Analyst extension. If your image turns solid black after you apply the tool, it means that the Interval tool does not work with that type of image data.

Identity Tool

Clicking on the Identity tool button will perform an identity transformation on the image, but not change any pixel values in the view. This limited tool works best with GRIDs showing categorical data such as land use or soils, but has no use for other image types.

Image Colormap

You can temporarily alter the color map of single-band images by using the Image Colormap. After clicking on the Colormap button, the Image Colormap dialog appears (see the following illustration), showing a colormap with 256 classes on the left, and additional control buttons on the right.

Image Colormap dialog.

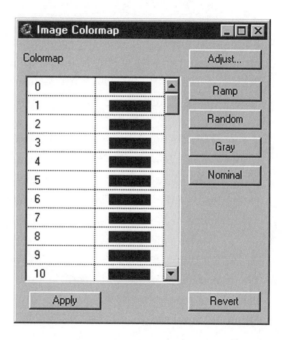

The Adjust button allows you to control the percent of red, green, and blue in the image, as well as the saturation and intensity of the image. For washed-out aerial photo images, try increasing the saturation using this button, in conjunction with upping the contrast with the Linear Lookup tool.

The Ramp button creates a color ramp from the color value assigned to class 0 to the color ramp assigned to class 255. This works similar to the Ramp feature for graduated color legends in the basic ArcView

Legend Editor. Double click on the color symbol to bring up the Fill palette for choosing your start color. Repeat the process to select your endpoint color, and then click on Ramp. The color ramp is automatically applied to the image.

> ✓ **TIP:** *To see more of your color ramp before applying it to the image, stretch the Colormap dialog box from the top or bottom to show more classes at once.*

The Gray button works in the same manner as the Ramp button, and automatically sets up a grayscale ramp and applies it to your image. Both the Ramp and Gray buttons are designed to work well with GRID images, but are less useful with aerial photos or scanned maps.

> ✓ **TIP:** *To obtain better relief features in a DEM, try the following procedure. For raw DEM GRID files, open the Colormap controls and select Gray and apply the feature. Then, increase or decrease the brightness of the GRID image by moving the line in the Linear Lookup graph to the right or left. This will produce a smoother depiction of elevation values, but will not produce a shaded relief.*

The Random button will assign a random color to each of the 256 classes in a single-band 8-bit image. The Nominal button will apply a nominal distribution of 16 color classes, repeated throughout the entire 256 classes to the image. This is useful when working with slope or aspect GRIDs.

> ✓ **TIP:** *Because the basic ArcView Image Legend Editor will not save changes to your legend, try using another image processing program to adjust your image. First, make sure to make a copy of your image and its world file (if any). Open the copy in a program such as Adobe PhotoShop and use brightness and contrast controls to sharpen the image. This will work as long as you do not resize or resample the image. Save the changes and add the copy of the image as a theme back into Arc-View. This process will not work with GRID files.*

Mapping with Common Image Types

Today there are a lot of free and low-cost images available that can make your mapping tasks much easier. Many of these images are downloadable from the web or offered on the Internet and provided on CD-ROM.

Using Digital Raster Graphics

Digital raster graphics, or DRGS, are scanned and geo-registered images of topographic maps produced by the U.S. Geological Survey. Several map scales are obtainable for free on line, including 1:250,000, 1:100,000, and the popular 1;24,000. Many companies provide value-added versions of DRGs, packaging all maps for an individual county or state on one CD-ROM, and removing the white borders around 7.5-minute maps. To find out were to download or purchase these products, go to one of the USGS DRG web pages at the following address.

http://mcmcweb.er.usgs.gov/drg/index.html

It may seem strange at first to use an image of a paper map in a digital map, but it has its advantages. There are two reasons to use DRGs in ArcView.

- As base maps from which to digitize data

- As backgrounds in digital maps with updated data themes placed over the image

DRGs are great to use as base layers from which to digitize details you do not have in other themes. You can take a paper topographic map into the field, record observations on it, and bring it back to the computer lab. You can then call up the corresponding DRG in Arc-View and then transfer the data from the paper map onto digital map as a new theme. This is a good way of producing 1:24,000-scale data that may not be obtainable anywhere else.

A free ArcView extension called DRG-Tools, produced by the California Department of Fish and Game, helps isolate features within standard digital raster graphics. Once isolated, different feature types can be used "as is" in a map, or used to provide a more focused base map for digitizing. The illustration that follows shows a standard DRG and a DRG with isolated contours.

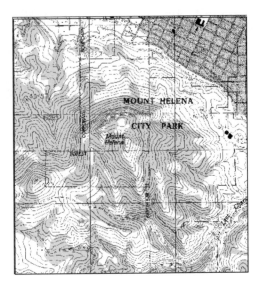 

Left image: A standard DRG. Right image: DRG with contours isolated.

Once loaded from the Extensions dialog in the File menu, DRG-Tools adds a new menu containing seven features in the view GUI. The features allow you to "Edgematch" or turn off the white paper border of standard DRGs, so that one map's border does not overlap another map's features.

Other options allow you to isolate different colored layers of the DRG, while making other colors temporarily invisible. This is a far preferable process for this type of image, rather than endlessly adjusting the brightness and contrast controls in the Linear Lookup tool in the Image Legend Editor. This is possible because the DRG is scanned as an indexed-color single-band image, and has a small number of basic colors, which correspond to features types on the topographic map. Table 10-3, which follows, summarizes DRG features and colors as used with ArcView.

Table 10-3: DRG Features and Colors

Feature	Color
Contours	Brown
Hydrography	Blue
Public Land Survey System and other survey grids	Red
Map updates	Purple and magenta
Miscellaneous features	Black
Vegetation	Green

Once isolated and hidden, all map features can be restored to the view using <Control+R>. Most of these features could be duplicated by cleverly using the ArcView Image Legend Editor, but it would be very time consuming to use the proper Colormap procedure. DRG-Tools combines all of these features in one place, and considerably speeds up the process.

DRGs add a lot of detail to an ArcView map layout, and add several items to watch out for. If you are using a DRG as a base and adding more themes on top of it, make sure to use line weights and colors that are different from those in the original DRG, to ensure that the overlying themes are legible. If you are using polygon themes, make sure to use the PostScript transparent hatch patterns in the *carto.avp* palette. These patterns will overprint the base map clearly.

Remember to check the metadata for each DRG, in that many of the 1:24,000-scale maps can be ten to thirty years old. In urban areas, many things can change on the ground over such a long period, and the DRG may only be useful for historical mapping purposes. You can zoom in to a DRG that still has its white-collared border and read the date of publication from the margin information.

Using several DRGs in a map can lead to printing problems. Try using ArcPress for ArcView, if you have it, or try a smaller portion of the DRG in your view. You can limit the displayed extent of an image in the View window, by going to Theme properties and selecting a specified extent, which is typed in map units in the dialog box.

> ↦ **NOTE:** *For an example of using DRGs in an ArcView map, see plate 11 of the color insert.*

Using Digital Orthoquads

Good aerial photos of a project area can be invaluable, providing up-to-date information of what is actually there on the ground. Good aerial photos are taken vertically with a mapping camera to minimize distortion of ground features, and then carefully geo-registered and rectified. Poor aerial photos can be an enormous time sink for the ArcView user, as you try to register an oblique image of a mountainous region to ground control points that just will not match up. The following illustration shows a digital orthoquad map with transparent hatched polygons.

Digital orthoquad map with transparent hatched polygons.

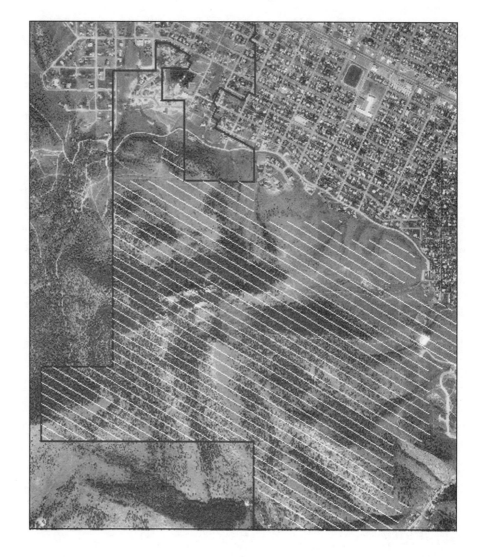

Rather than wrestling with poor aerial photos, consider using the Digital Ortho Quarter Quads, called DOQQs, which are being produced by several U.S. government agencies for most of the country. DOQQs are high-quality aerial photos precisely registered to a 1;24,000-scale topographic quad sheet. The DOQQs meet national map accuracy standards and have an extraordinary resolution of 1 meter. Because each image contains so much detail, each quad sheet is broken up in quarters, hence the name *quarter quads*.

> ➥ **NOTE:** *For more information on obtaining DOQQs visit the USGS DOQ web page at the following address.*
>
> http://mapping.usgs.gov/mac/isb/pubs/factsheets/fs03900.html

DOQQs are saved as single-band grayscale images in a couple of different file formats, including the *geotiff* format. The contrast of the grayscale images can be adjusted using the Linear Lookup tool in the Image Legend Editor, in order to highlight images features and increase legibility of overlying themes. For text, try using bold typefaces filled with white to increase contrast with the darker gray background of the image. Transparent hatch patterns for polygon themes from the *carto.avp* palette work well with DOQQs, as well as with DRGs.

Each quarter quad can be as large as 150 megabytes in file size, making a view of a full quad sheet in the 400- to 500-megabyte range. Obviously, you will not be printing dozens of these DOQQs on one map anytime soon! ArcPress is a must for faster printing.

> ➥ **NOTE:** *For an example of using DOQQs in an ArcView map, see plate 12 of the color insert.*

Summary

Imagery adds a powerful tool to many maps, and basic ArcView can use dozens of popular image formats in the map production process. To take full advantage of raster data, ArcView users need specialized extensions, including ArcView Spatial Analyst, to manipulate this type of information.

∞ **NOTE:** *For more techniques on using shaded relief and 3D imagery in your maps, see the web site for* Cartographic Design with Arc-View GIS *at the following address.*

http://www.onwordpress.com/olcs/index.html

Chapter 11

Map Output

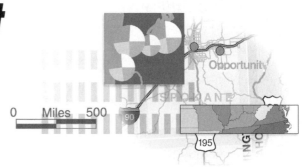

Printing technology in the last decade underwent a staggering change, almost equal to the rate of change in personal computers. Cartographic products used to be limited to simple gray-scale maps, clunky pen plotters, and very expensive color publishing. Today there are a bewildering variety of choices for print output. Short-run color is common, but what type of color do you need? The choices range from simple four-color prints, to stunning photorealistic prints from $200 personal ink-jet printers, to large, high-resolution posters generated from specialized plotters.

You now have the option of not printing your map at all, but posting it on the web for screen viewing or downloading. Then there is the developing field of interactive maps on the web, where maps can be customized, queried, and analyzed, all within a web browser.

Most of these methods of map output require different design considerations. If you have planned ahead, you know what map product your audience needs and have worked any special design considerations into the mapmaking process. Now it is time to get it on paper (or on screen if you are creating output for the web).

Global Considerations

Just as important as knowing what the audience is for a map before you begin making it is knowing what type of printer will produce your final hardcopy map. There are several global considerations to keep in mind when planning to print your map.

Different printers, even of the same type, can produce quite different products. A basic assumption in the descriptions that follow is that your computer with ArcView is connected directly to the printer, or connected to the printer via your office network. You will have to use ArcView's file export feature if you need to send map layouts via the Internet or by disk to a remote printer.

Color and Grayscale

An obvious difference is of course color versus grayscale printing capability. Most recent grayscale laser printers produce prints with resolutions of 600 dpi (dots per inch) or higher. Therefore, variations among similar black-and-white ArcView maps produced with these machines will be minor. There are many types of color printers, from simple color inkjets, to color lasers, to arcane dye-sublimation printers and color plotters—all of which can produce strikingly different variations in the output of the same map file. The idiosyncrasies of color and grayscale printing are further developed elsewhere in this chapter.

Page Size

Page size is another obvious factor. Most printers, both grayscale and color, can produce letter- and legal-size maps. Many newer inkjet printers and a few laser printers can print out tabloid (11 inches by 17 inches) copies, sometimes called ledger size. Tabloid-size maps are very useful for color inserts to documents, because they span two letter-size pages, and can be simply folded into a report.

Anything larger than tabloid size is poster size. Such maps are produced from plotters, common in cartographic and engineering firms. These are also made using high-resolution poster image-setters found in advertising shops or printer service bureaus.

Margin Size

All of these printers can have different margin settings. Printers can commonly print out to 1/4 or 1/2 inch from the paper's edge. A few specialized printers can produce a full bleed; that is, print color all the way out to the edge of the sheet of paper.

The margin size is normally reflected as a thin blue line, just inside the outline of the page size in an ArcView layout. The position of this thin blue line will change if you change printers in the Print Setup option found in the layout's File menu. ArcView layout margin size options are shown in the illustration that follows.

✗ **WARNING:** *Do not trust this little blue line! ArcView's margins do not always accurately reflect the printer's margins, especially for maps larger than letter size, and map features placed near the blue line may be cut off when printed. Make sure to do a draft print before placing features near a page's margins.*

Margin size options in the ArcView layout.

✓ **TIP:** *When you are producing maps to be bound directly in a document or report, remember to leave an extra-wide margin (at least 1/4 inch, and possibly as much as 1.5 inches) on one side for binding or stapling. Do this on the left margin if the document is to be bound right-reading, or on the top margin if the map will be bound in the document as a landscape image.*

Printing Time per Copy

After furiously working for hours, or even days, on a complex map in ArcView while trying to meet a deadline, it can be cause for panic to see the map file disappear into the printer and remain there, as the printer's busy light blinks and blinks. Printers can take anywhere from a couple of minutes to many hours to correctly image a map.

Printer speed is dependent on the amount of RAM (memory) in a printer, whether the printer has a fast CPU on board or not, and the complexity of the map file. ArcView layouts containing a lot of line features and pattern fills, thousands of pieces of type, or complex images can take quite a while to print. Aside from simplifying your map layout or purchasing a faster printer to save time, you have few options here.

Changing the resolution setting in Page Setup can save printing time for layouts containing raster images such as digital elevation models, satellite images, or any type of photography. Go to Page Setup under the Layout menu and examine the Output Resolution feature. Resolution can be set to Low, Normal (the default), or High.

Changing the resolution makes no difference when printing vector graphics, but can make a big difference in printing time regarding images. High resolution utilizes all the pixels in an image when printing, which can result in a large file in the printer. Normal resolution re-samples the image and utilizes one out of every two pixels. Low resolution re-samples the image and uses one out of every four pixels. The following illustration shows ArcView layout image resolution options.

Image resolution options in the ArcView layout.

Low = 1/4 of the pixels in an image

Normal = 1/2 of the pixels in an image

High = All the pixels in an image

Your printed results depend on the area occupied by the image on the page, and the quality of the image at the start. If your image takes up a comparatively small portion of the page, such as a two-inch-square area, the low-resolution setting may not be noticeable on the final print.

Conversely, if the image takes up the majority of the layout page and the quality is not high to begin with, using the low or normal resolution settings may produce an obviously "grainy" or "bitmapped" print. To successfully print maps with large images, you may need to use a software raster image processor (RIP) such as ArcPress. (RIPs are discussed in more detail in material to follow.)

Degree of PostScript Support

For the highest quality of map output, you need a printer that supports the Adobe PostScript page description language. PostScript is an object-oriented programming language that uses patented algorithms to describe all types of graphics, fonts, and colors. Because features are described mathematically, they can be printed at any page size on any PostScript printer at any resolution, and retain their shape and clarity. As a result, PostScript graphics are called resolution-independent graphics.

PostScript has evolved over the years in releases called levels, and ArcView's printer drivers have changed to take advantage of PostScript's new features. PostScript Level I was in common use from 1985 through 1991, but is not used today (in 2000), except by older printers sold during that time period. ArcView's Basic printer driver supports PostScript Level I. PostScript's printer driver options are shown in the illustration that follows.

→ **NOTE:** *If your computer is connected to an office network that uses a print spooler on another computer, you might not have the PostScript driver options available in the ArcView Print dialog box. ArcView should default to the Enhanced PostScript Level II driver in this case. If you have problems with printing, try printing your layout directly from the print spooling computer, or convert the layout to an Adobe Portable Documents File (PDF), and print that file from Adobe Acrobat.*

*PostScript
printer driver
options.*

PostScript Level II was released in 1992, and was quickly incorporated into printing technology. ArcView's Enhanced printer driver takes advantage of PostScript Level II printing, but will not print when used with an older PostScript Level I printer.

PostScript Level III was created in 1997 and is now offered in many printers on the market. Level III includes many new features for printing, including very fine, smooth shading of grays and color blends, with no banding even at high resolutions at thousands of dpi; precise color controls for several different color models; improved image masking; and much faster printing of complex graphics (such as maps!). ArcView 3.2 does not have a PostScript Level III driver as yet, and users may have to wait for the ArcView 4.0 release for this built-in capability.

> ✓ **TIP:** *If you have a new PostScript printer or plotter with Level III, try using the ArcView Enhanced printer driver. PostScript Level III printers can automatically interpret older versions of the PostScript language.*

To use PostScript Level III for complex maps, you may be able to print the map layout from ArcView 3.2 as an Adobe Portable Document File (PDF), and then print the map from Adobe Acrobat Reader to the new printer. Alternatively, take advantage of the Epson Stylus RIP for Epson inkjets, which utilizes PostScript Level III technology.

PostScript RIPs

PostScript is capable of creating very high-quality maps and graphics because it represents points, lines, polygons, and images in the computer as mathematical equations. However, nearly all printers—whether lasers, inkjets, or plotters—do not print equations. Printers represent graphics on paper as a series of tiny dots. To go from equations in the computer to dots on paper, you need an RIP. RIPs are of two types: hardware RIPs and software RIPs.

Hardware RIPs

The most common type of hardware RIP you are likely to encounter is the interpreter built into many higher-quality PostScript laser printers and many plotters. These RIPs are essentially invisible, doing their job converting PostScript code into dots inside the printer. Printers that have internal PostScript RIPs (also called PostScript boards or PostScript interpreters) have to have a CPU on board as well to perform the calculations. This process is frequently referred to as ripping the print file.

One advantage of having the print file ripped inside the printer is convenience. Once you activate Print from the File menu in an Arc-View layout, ArcView creates a PostScript file and dumps it into the printer. You are free to do something else with your computer while the printer's RIP conveniently handles the rest of the printing work.

The addition of PostScript capability, along with a hefty CPU, results in paying significantly higher prices for a PostScript printer or plotter, running anywhere from $500 to several thousand dollars more per machine. Many machines will come with PostScript built in at the factory, but some can have the capability included later, by the installation of a PostScript add-on board.

Much less common than internal hardware RIPs are external RIPs. Several companies (such as Birmy, Heidelberg, and Electronics for Imaging [EFI]) produce high-end external PostScript RIPs (loaded with RAM memory and lightning-fast CPUs), which can be added to plotters, image setters, or color copiers.

You are likely to encounter one of these at the more sophisticated print shops or advertising agencies that offer short-run color publishing or very high-resolution color poster printing, such as Iris prints. Iris prints are digital prints created from electronic file output as near-continuous tone images with a resolution of 1840 dpi on virtually any substrate up to 35 inches by 47 inches. Iris prints appear as photorealistic- or lithographic-quality output. It is unlikely that ArcView would be found on a computer hooked up to one of these exotic techno-beasts, but you can print fine maps from them by saving your ArcView layout as an Adobe PDF document and taking that file to the print shop.

Software RIPs

As maps and graphics have gotten more complicated and more difficult to print, technology companies have developed software RIPs to handle the problem. A software RIP is a stand-alone computer program that grabs PostScript files generated from other programs, such as ArcView, and rips the files into the native language of the type of printer you are using. A software RIP takes over the job that would be done by an internal PostScript hardware RIP in the printer.

Software RIPs have many advantages over (and some disadvantages compared to) hardware RIPs, the first of which is that they are usually less expensive, ranging from $100 to $400 for the stand-alone programs. Given enough time, software RIPs can handle incredibly complicated map files, such as PostScript files that would fail to print on all but the most sophisticated hardware RIPs.

> ✓ **TIP:** *Before buying a software RIP, or buying a new printer for that matter, make sure a software RIP will work with the type of printer you will be using. Some software RIPs, such as the Epson Stylus RIP or Birmy PowerRIP, only work with a limited number of printers. Arc-Press for ArcView works with a couple of dozen printers. As new printers come out and the RIP capabilities change, your printer might not be among those compatible with the software. Always check with the software manufacturer before committing to a new printing setup.*

Software RIPs usually run on your local computer, and a file in process can lock up your machine's CPU and memory resources for considerable periods, from ten minutes to several hours. Color maps run through a software RIP tend to have a big color shift, with lighter colors becoming darker and darker colors over-saturating. Once again, what you see on the screen is not necessarily what you get on paper.

ArcPress for ArcView

ESRI has recognized the problem of printing complicated maps, and has developed ArcPress 2.0 for ArcView. ArcPress is a stand-alone PostScript Level II software RIP that also acts as an extension from within ArcView. Presently, ArcPress works with two dozen different printers produced by seven manufacturers. Table 11-1, which follows, lists the printer file types ArcPress will rasterize.

Table 11-1: ArcPress for ArcView-supported Printer File Types

Manufacturer/Type	Printer Format	Notes
Calcomp Electrostatic	CCRF	No longer manufactured.
Calcomp Techjets	CCRF-IL	Supports all Techjets.
Canon Bubblejets	CBJ, RIC (Raster Image Commands)	Supports most Bubblejets.
Encad Novajets	RTL (Raster Transfer Language)	Supports all Novajets.
Epson Stylus Pro Inkjets	ESC/P2	Supports Stylus Pro, Pro XL, and Stylus 1500.
Hewlett-Packard Laserjets	PCL3-PCL5	Supports HP Color Laserjet.
Hewlett-Packard Deskjets	PCL3-PCL5	Supports all Deskjets.
Hewlett-Packard Designjets	RTL (Raster Transfer Language)	Supports all Designjets.
Raster Graphics	RGI	Supports all raster graphics.
Versatec Electrostatic	VRF	Supports Versatec Electrostatic.

✓ **TIP:** *If you work in a networked office environment that includes ArcInfo and ArcPress for ArcInfo, you may not need ArcPress for ArcView. Try the sample extension* aiapress.avx *from the ESRI ArcScripts web page to access ArcPress for ArcInfo as an alternative to purchasing ArcPress for ArcView.*

ArcPress for ArcView, currently in version 2.0, loads from the Extensions option in the ArcView File menu. It becomes available as another option in the layout's File menu, and also appears as a button on the right of the layout's button bar. When your map is finished, you can click on the ArcPress button and the layout will be exported to the ArcPress extension as a PostScript New (EPS) export file. For large map layouts on slower computers, this export process can take anywhere from five minutes to half an hour.

Once the EPS export is complete, the first ArcPress dialog box pops up in the center of the screen. Here you need to choose where you want ArcPress to place the temporary EPS file it will try to rasterize, normally the *C:\temp* directory. ArcPress for ArcView is shown in the following illustration.

ArcPress for ArcView.

Make sure you have a great deal of space available on your selected hard drive for these temp files! ArcPress needs a lot of breathing room to create both the temporary EPS and native printer language files it is about to make. Very complex maps can take 100 to 200 megabytes or more of temp space, and can lock up your computer if the space is not there when it is needed. You can examine the EPS file statistics and size from the File Info button, but do not try to skimp on free space.

In addition, ArcPress does not clean up after itself. After ArcPressing several files successfully to a printer, go back and toss the temporary EPS files in the trash to free up hard drive space, or rename them and save them to another directory if you need to reprint the same map soon.

At this point, you have the option of rasterizing the map file and exporting it to a number of common raster formats. You may want to do this if you want to import the file into another graphics or image processing program. Table 11-2, which follows, lists the raster image file export types.

Table 11-2: ArcPress for ArcView Raster File Export Options

File Type	Definition
TIFF	Tagged Information File Format
JPEG	Joint Photographic Experts Group
PNG	Portable Network Graphics
PCX	PC Paintbrush file format
BMP	Microsoft Bitmap format
BIP, BIL, BSQ	Band Interleave file formats
PBM, PGM, PPM	Portable Bitmap formats

More often than exporting files out of ArcPress, you will continue by pressing the print button to bring up the ArcPress Print dialog box, shown in the following illustration. The first task here is to match the printer type to the device type. The Printer drop-down menu lists the printers attached to your computer or your office network. The Device drop-down menu lists the native printer languages ArcPress can rip the EPS file into.

ArcPress for ArcView Print dialog box.

If you chose a Canon BJC Inkjet printer, you need to select Canon Bubblejet (RIC) device. If you chose an HP Designjet plotter as the printer, you need to select the HP RTL language device. Printers will accept only their native software device language, and not that of another printer manufacturer.

Before clicking on OK, you have the choice of setting page layout and color options, by clicking on the Options button. In the Page Layout Options tab, you can set margins, crop the extent of the print file, scale the file from 25 to 200%, and rotate the file. You can also attach a handy custom banner with printer names, and dates and times; embed trim guides around the map; and set tiling options for large maps. The tiling option allows those of us with letter-size print-ers to construct larger-than-letter-size maps. The page layout options are shown in the following illustration.

ArcPress for ArcView Page Layout options.

The Color tab under Page Layout gives you a lot of control over the way color is finally printed on paper. Remember here that you are taking an EPS file with CMYK colors, whose colors are chosen as HSV colors in a view in ArcView, and translating that file into a particular printer's software language.

You are likely to obtain a color different from the colors you can see on the map on the screen, or from the colors you would get if you printed the EPS file to a PostScript printer. If you are using ArcPress for the first time, do not play with the Color options, but go back and just print, to see how your map comes out. The ArcPress color options are shown in the following illustration.

ArcPress for ArcView Color options.

The color options allow corrections of individual CMYK colors, adjustment of overall color saturation on the map, changes in the color configuration (rarely needed), and selection of an alternate dithering process. ArcPress color outputs are frequently darker and more saturated than EPS files printed without ArcPress, sometimes to the point of obscuring text or other map details. First try lowering the percent of individual colors. If that fails, try altering the total saturation of the map.

Maps printed on some printers that do not support true color may have pronounced banding or streaking. This can be avoided by changing the dithering pattern in the Color options. Dithering adds random noise to the final image and gives the illusion of a wider range of colors, which will prevent banding. Most newer printers and plotters, however, support true-color output, making dithering unnecessary.

Once you have fussed with the options and ArcPressed enough maps to settle on something that prints to your satisfaction, you can save the options file to load and use later for similar types of maps. For example, if you are ArcPressing a series of maps utilizing color aerial photos or satellite images of a region, you would want to save an options file to correct the color changes for that entire series of maps.

> ✓ **TIP:** *ArcPressing to some types of printers may have a few quirks you will need to work around. The file may have to be rotated 90 degrees to save paper and to print without clipping. ArcPressing may not work at all if a nesting option is turned on for some HP plotters, resulting in large sheets of blank paper gushing out of the printer! Talk to your printer manufacturer, or post a question on an online help list such as ArcView-L to get answers to your printer-specific problems.*

By finally pressing the print button, ArcPress starts the process of interpreting the EPS file into the printer's native software language. Again, this might take a minute to half an hour, depending on the file size. Once the interpretation is complete, the map will quickly print on paper.

Other Software RIPs

A handful of companies produce PostScript Software RIPs. The three most popular are discussed here. These software programs differ from ArcPress for ArcView in two main ways. Unlike ArcPress, they are designed to work with the output of a variety of software programs, not just ArcView.

Also, unlike ArcView ArcPress, which works with a variety of printers, the Birmy PowerRIP and the Epson Stylus RIP are brand specific, and will not work with printers from different manufacturers. If you are producing graphics or graphic-rich documents from other programs in addition to maps from ArcView, these ripping programs deserve a closer look.

Birmy Graphics PowerRIP 4.0

Birmy Graphics Corporation *(http://www.birmy.com)* produces the PowerRIP for Canon Inkjet printers. Canon was the first company to offer an 11-inch by 17-inch format inkjet printer, the BJC-4550, which was specifically aimed at mapmaking from GIS and CAD programs. The PowerRIP, whose Control Panel is shown in the following illustration, sells for about $125, works as a "virtual printer" accessible from the Windows or Macintosh desktop, and supports PostScript Level II and Adobe brilliant screening.

Birmy PowerRIP.

Epson Stylus RIP

Epson Corporation *(http://www.epson.com)* broke new ground by offering the first PostScript Level III software RIP in late 1999. The Stylus RIP works for the popular Epson line of Stylus inkjet printers, and sells for about $100. The advanced PostScript Level III capability is provided for Windows NT 4.0 and Macintosh platforms, and PostScript Level II is offered for Windows 95 and 98 systems. In addition, Epson Stylus RIP offers Pantone Color Matching capability, a valuable feature if you are going to produce a lot of maps using spot colors for publications.

Image Alchemy PS

Image Alchemy PS from Handmade Software *(http://www.hand-madesw.com)* is the current version of one of the original software RIPs ever produced. This venerable program won (and still has) a small but dedicated following for its cross PC-UNIX platform functionality, its ability to process over 90 different graphic formats, and its DOS-like command line interface.

Image Alchemy PS offers a PostScript Level II RIP on Windows or UNIX platforms and retails for about $300. Many GIS professionals used earlier versions of Image Alchemy with ArcView and ArcInfo before the arrival of ArcPress.

The Printing Process

Printing maps can be a complicated process, and it is worthwhile to understand the ins and outs of how it works. The starting point is always a finished map in an ArcView layout. You have set your page size and orientation in the Page Setup feature of the Layout menu, and you are ready to print.

In the Print Setup dialog box under the File menu, you can choose your printer and set printer-specific options from the Properties button. Make sure to check the paper size and orientation again, because the properties that are set under Page Setup in the Layout menu do not automatically translate to the properties in the Print Setup dialog box.

After clicking on OK, go to Print in the File menu. In the Print dialog box, the Setup button gives you another access to Print Setup features, and the Print to File check-off box permits you to save the layout as a *.prt* file on your computer's hard drive.

✓ **TIP:** *PRT files created with the Print to File option cannot be reopened with ArcView, but they can be printed using the command prompt. In Windows, open a command prompt window and use the Copy command to copy the .prt file to your local printer. For example, if your print file is on your D drive and is named mapone.prt and you want to send it to a printer attached to the LPT1 port on your computer, you would type the following.*

```
Copy d:\mapone1.prt lpt1
```

The *.prt* file would print on your local printer. If you want to save ArcView layouts as external files, consider using Adobe PDF files rather than using the Print to File feature.

Printer Drivers

If your printer is connected directly to your printer, and not accessed over the network, the Print dialog box will offer a choice of three radio button options under the PostScript Interface. The ArcView Enhanced driver produces PostScript Level II output and will print to either PostScript Level II or newer Level III printers. This is the driver of choice for best graphic output for many printing situations. The ArcView Basic driver produces PostScript Level I output for older (early 1990s) printers.

The Native OS driver will use either a printer-specific driver, normally offered by the printer's manufacturer, or a Microsoft driver if you are using a Windows PC. The printer-specific driver can be the most efficient means of printing to many types of inkjet printers or larger plotters. These hardware-specific drivers usually come on disk or CD when you buy your printer, but are frequently updated as computer operating systems change. For best results, check the printer manufacturer's web site for newer versions of your printer driver.

Spooling and Printing

Once you click on OK in the Print dialog box, ArcView begins to create a file from the layout to send to the printer. The file may be spooled on your computer locally, sent over the network to a print spooler, or downloaded into the printer and spooled on a remote hard drive.

Where the spooling occurs can make a huge difference in printing time and performance for large map files. Modern plotters have large hard drives, which function as spoolers built into the plotter, relieving the network of much of the printing process. Large GIS shops may have one computer dedicated solely as a network spooler and ArcPress RIP, making the printing process even faster and more efficient.

Once the print file arrives at the printer, the printer's CPU begins building a complete picture or raster image of the map in the printer's memory. The printer uses information in this picture to begin spraying liquid ink or setting dry toner onto the paper. If the printer does not have enough memory to handle a large file, it may fail to print or just print a portion of the map.

If you are printing to a printer without RAM memory or a CPU, like most inkjets, the process of creating the "picture" of the map occurs in your local computer's memory, which is tricked into serving as the printer's CPU. This may slow down other processes on your local PC, depending on the size and complexity of the print job.

With ArcPress, the computer's or printer's CPU does not have to create the entire picture of the map before it begins to print. ArcPress will send several lines of the raster image of the map to the printer, wait for them to be placed on paper, and then send several more lines. This incremental process prevents overloading the printer with a large map file.

Types of Printers

There are a bewildering number of types of printers available today, giving you a bewildering number of options for getting your ArcView map on paper or in print. As always, it is good to keep in mind what map product you need to satisfy your users, and what printing technology you have access to for producing the proper map product. The sections that follow discuss various types of printers.

Grayscale Lasers

Laser printers that produce grayscale or black-and-white output are the most common type of printers found in offices today, and range in price from $500 to several thousand dollars. These workhorses are chosen for their ability to print large documents consisting mostly of text on paper.

Grayscale lasers use a laser beamed onto a photosensitive cylinder, which attracts black toner from a cartridge. Print size is limited to letter or legal, using primarily bond paper. Many later-model laser printers print at 600 dpi resolution or higher, and frequently come with PostScript capability built in. As a result, graphics print crisply in grayscale, and typography is clear and distinct on the map.

Map output options in grayscale, however, are limited. Without using pattern fills, it is difficult to portray more than four or five feature types using shades of gray the map reader can discern. Color maps sent to grayscale lasers will rarely have the same impact they have when printed in color. Darker colors print almost black, and lighter colors all blend into the same level of gray.

> ✓ **TIP:** *If you are limited to grayscale laser output, design your map in ArcView using levels of gray, not color, for best results.*

To ensure high-quality shades of gray from laser printers, make sure the internal workings of the printer are clean and are free from paper dust. Also make sure the toner cartridge is fresh and not near the end of its life span.

When using pattern polygon fills instead of solid gray in ArcView, make sure to use the *carto.avp* fill palette that contains PostScript pattern fills. If you do not, you will notice the difference when the pattern files print as jagged bitmaps or dark solid fills on a PostScript laser, as shown in the following illustration.

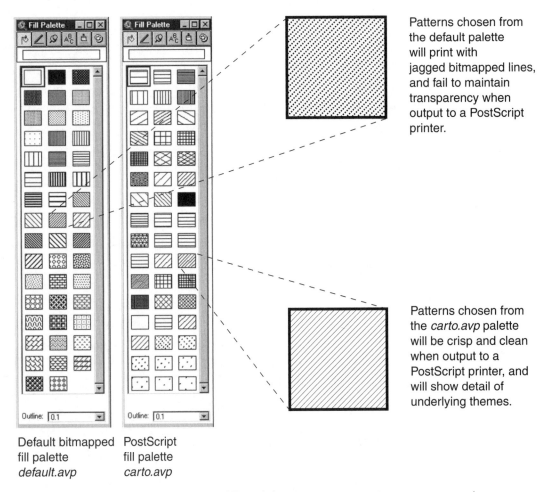

Patterns chosen from the default palette will print with jagged bitmapped lines, and fail to maintain transparency when output to a PostScript printer.

Patterns chosen from the *carto.avp* palette will be crisp and clean when output to a PostScript printer, and will show detail of underlying themes.

Default bitmapped fill palette
default.avp

PostScript fill palette
carto.avp

A comparison of the default pattern fill and the carto.avp *PostScript pattern palette.*

Color Lasers

Less common than grayscale lasers and more expensive are color lasers. These pricey machines range from $1,500 to almost four thousand dollars, but work in basically the same manner as their less colorful cousins. Instead of one toner cartridge, color lasers have four to make up the standard cyan, magenta, yellow, and black (CMYK) color model.

In order to manage color, these lasers come standard with PostScript capability, and usually loaded with 30 megabytes or more of RAM memory. Still, the demands of high-resolution (600 to 1,200 dpi) color printing can make these printers slow and cumbersome when it comes to printing complex ArcView maps. Color, which is generated by mixing the four types of dry toner, tends to be more saturated and less bright than color produced by inkjet printers.

Color Inkjet Printers

The technology behind color inkjets made enormous strides during the 1990s. Short-run color printing, which used to be expensive and slow, today is relatively fast and inexpensive. Quality color inkjets can be found at any office supply store for as little as $100, with top-flight off-the-line inkjets selling for $400 to $500.

Inkjets use liquid inks in CMYK colors, which may be contained in one or many ink cartridges. Microscopic amounts of individual colors are heated to high temperatures by small electrodes and sprayed out onto paper at resolutions from 300 to 1,000 dpi. Text printed in black by many inkjets rivals the quality of lasers, while offering full color to boot. A tabloid-size color inkjet printer is shown in the following illustration.

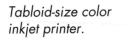

*Tabloid-size color
inkjet printer.*

Some inkjets print to 11-inch by 17-inch paper, a great size for maps, whereas others produce photo-quality images, a great advantage for shaded relief maps or satellite images. One reason that inkjets are so inexpensive is that few of them contain a CPU, like most laser printers or plotters. Your computer's CPU must do the heavy lifting to interpret the ArcView layout into the inkjet's native language, and produce the map on paper.

Most inkjets contain only a megabyte or less of on-board RAM memory, primarily to act as a buffer that holds portions of the print file as it is transferred from a PC. This means that printing more complicated maps from ArcView on an inkjet printer can be very slow and time consuming.

Although an inkjet printer may be rated at a page per minute for printing text with simple graphics, a moderately complicated map may take 10 minutes to half an hour or longer. Clever mapmakers work around this limitation by printing a draft copy of the map during normal work hours, proofing and updating the copy, and then running multiple final copies overnight. For even more complicated maps or just to speed up day-to-day workflow, it may be worthwhile purchasing an RIP.

Although the initial cost of a color inkjet printer may be low, supply costs can add significantly to day-to-day maintenance expenses. Ink generally comes in small cartridges, which run anywhere from $20 to $60 per cartridge. Simple letter-size color maps may cost only 10 to 20 cents per page for ink, whereas complex maps and especially photographs may run the price up to several dollars per page. With the high cost of cartridges, it is always good to check if they are refillable; some are and some are not. Third-party dealers provide several refill options for Canon, Epson, and HP inkjet cartridges.

Color Plotters

Plotter technology has changed dramatically in the last decade, emulating many of the improvements seen in smaller inkjet printers. Older plotters are of two types rarely seen in use anymore: pen plotters and electrostatic plotters. Pen plotters used four pens that drew lines of different colors on paper.

Although good enough for blueprint or simple CAD-type drawings, pen plotters could not draw solid fills and could not draw images. Electrostatic plotters worked as large, crude laser-writers, and used electrical fields to place colored ink on paper. Both of these expensive technologies have been pushed aside by improvements with cheaper inkjet printing.

Popularized by Hewlett-Packard under their Designjet line and copied by many others, large-format inkjet plotters work on the same basic principles as their smaller cousins. Minute quantities of colored ink (CMYK colors) are heated to high temperatures and precisely sprayed onto the paper. Current inkjet plotters can reach print resolutions of 600 dpi or higher, and produce near-photo-quality images. An example of such a plotter is shown in the following illustration.

*HP 1050CM Designjet
PostScript Level III
plotter.*

Inkjet plotters differ from their smaller cousins in significant ways, including the ability to handle larger paper sizes. Most inkjet plotters have a CPU on board to handle the ripping process. Fully loaded inkjet plotters can have 128 megabytes of RAM memory and multi-gigabyte hard drives, allowing them to rip enormous print files with ease. Some inkjet plotters, such as the HP Designjet 1050CM, are currently shipping with the new PostScript Level III capability as well.

ArcView works very well with this new generation of inkjet plotters, transferring print files as PostScript or HP RTL language from the PC into the plotter's RAM memory or onto the plotter's hard drive. Multiple copies of maps can be "tiled" or "nested" onto larger sheets to avoid wasting paper, and transparent Mylar or waterproof poster paper can be run through the plotter.

One consideration of printing to be taken into account with inkjets of all sizes is the tendency of maps to fade with time. It is common for map users to request another copy of an older map and be surprised at the "different" color scheme than the one presently hanging on their wall.

Inkjet ink is light sensitive, and will fade if exposed to direct sunlight over a period of weeks or months. Maps can be laminated in satin or semi-matte film, which will prevent this fading, and make the map resistant to tearing and waterproof for use in the field. Laminating a dozen copies of a small map that will be handled a lot at meetings or briefings is a nice professional touch for your final map product. Laminating or sealing is a service found in many of the more modern print shops in urban areas.

Imagesetters

Imagesetters are high-resolution printers used in the publishing and prepress industry. As a mapmaker, you will run into this technology anytime you submit a map for publication in a book, atlas, or magazine. Imagesetters take PostScript files, separate them into different colors, and image them at thousands of dots per inch resolution onto film or directly onto printing plates. The film is used to make printing plates, which hold the colored ink as the printing press moves paper over them.

ArcView, unlike ArcInfo, does not have a color separation utility. To prepare your map for publishing, you have two options: export the map as a PostScript (EPS) file from the Export option in the ArcView layout, or print the map to an Adobe PDF file if you have PDFwriter. Either file type can be delivered by disk or the Internet to an imagesetter at a service bureau or publishing house.

Professionally publishing a map is an involved and complicated process. If you are unfamiliar with the process, make sure to talk to an experienced digital graphic designer and your publisher before proceeding too far. They can provide advice on color and font selection, as well file formats. Publication maps are frequently finished in a dedicated graphics program, such as Adobe Illustrator or MacroMedia Freehand. Familiarity with these software programs and their color separation features is a great help in the publishing process.

Printing Tips

There are a host of things you can do to streamline the printing process, and avoid those last-minute deadline crises that seem to occur near the end of the map production process! Use the tips that follow to plan ahead and you will be ready for most map printing jobs.

Printer Calibration

Calibrate your printer. Take some time to know the capabilities of your printing hardware. Print the *calibrat.apr* layout, found in the ArcView samples directory, on all of your printers, and compare their colors to those of your computer's monitor. Use the Symbol dump script (*symdump.ave*) to print layouts of your most popular palettes, including colors, point markers, and pattern fills.

Try loading and printing the *artist.avp* palette to give a wider range of colors in a mapping project. Save these printouts in a binder, and refer to them before getting too far along in the process of constructing a map. This will prevent the awful last-minute surprise of finding out a half hour before your important presentation that the cute symbol you chose to use all over the map resembles some sort of squashed insect when finally printed.

Printing Materials

Know your paper and other printing materials. Keep a good supply of the various types of paper that work well with your printer on hand. High-quality bond paper can make all the difference when printing a complex color map. Try out the cheaper discount papers on your printer well before you need to print that critical job.

Some printers, especially photo-quality printers, require special coated paper to hold the colors correctly. Using cheaper materials can result in blurry details and ink smearing over the map. When printing overhead transparencies, check your printer's specifications beforehand to see what brands of the clear plastic overheads can be run through them, and avoid the expensive embarrassment of having the wrong ones melt inside your new $4,000 color laser.

If you are using ArcPress, have your ArcPress options tested and saved to avoid the common problem of over-saturating your prints. Run the *calibrat.apr* layout and your palettes through ArcPress and compare the printouts to those done without ArcPress.

Printing Help

Know how to get printing help. Investigate local print or copy shops to see what types of high-volume printers or color copiers they have available. Print one high-resolution print of your map on your office printer, and take it down to the copy shop for 100 color copies. If the print shop has a high-end color printer, bring down an EPS export from ArcView or a PDF of your map and print final copies on the higher-quality machine.

Exports to Other Programs

When printing maps directly from the layout document will not suffice, ArcView offers the ability to export layouts to ten different file types for use in other programs. Which export format you choose is dependent on what you want to do with the file in the other program. Some export files allow you to edit them in other programs, whereas others are better only for placement purposes.

To access the Export features, go to Export under the File menu when in the layout or the view document. When the Export dialog box appears, select your intended file type from one of the ten options under List Files of Type in the drop-down menu in the lower left corner. Before clicking on OK, check out the features in the Options button on the right of the Export dialog box. Name your export file and click on OK, and ArcView will attach the appropriate suffix for it after it is saved to the hard drive. The Export features are shown in the following illustration. Table 11-3, which follows, lists the ArcView raster and vector file export options.

Export feature and export options dialog box.

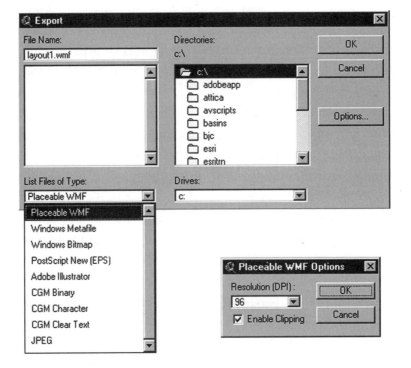

Table 11-3: ArcView Raster and Vector File Export Options

File Type	Definition	File Suffix	Raster or Vector
Placeable WMF	Placeable Windows Metafile	.wmf	R & V
Windows Metafile	Windows Metafile	.wmf	R & V
Windows Bitmap	BMP Microsoft Bitmap format	.bmp	R
PostScript New (EPS)	PostScript Level II file format	.eps	R & V
PostScript (EPS)	PostScript Level I file format	.eps	V
Adobe Illustrator	Adobe Illustrator PostScript format	.ai	V
CGM Binary	Computer Graphics Metafile	.cgm	R & V
CGM Character	Computer graphics characters	.cgm	V
CGM Clear Text	Computer graphics clear text	.cgm	V
JPEG	Joint Photographic Experts Group format	.jpg	R
PICT	Picture format	.pct	R

Placeable WMF and Windows Metafile Exports

Both the Placeable Windows Metafile and Windows Metafile formats are commonly used formats easily manipulated in Microsoft Windows Office programs, as well as in Adobe Illustrator, Corel Draw, and Hijack Graphics. In Microsoft Office programs, the resulting vector graphic can be resized, fills and line weights changed, and text edited.

Under the Options feature, you can choose a resolution for any raster images included in the layout, at 72, 96, 120, and 144 dpi. As you might guess, the higher resolutions result in much larger file sizes than those of lower dpi, but will print more crisply. The Enable Clipping check box does not seem to do much of anything to the final export file, and its function remains known only to the programmers at ESRI!

✓ **TIP:** *Saving layouts as Windows Metafiles is a fine way of including ArcView maps in large text documents, such as reports or briefing* *books. WMF files can also be popped right into Microsoft PowerPoint and used for quick presentations at meetings or conferences. You can save archives of ArcView maps as WMFs in PowerPoint and recall them when you want to review past projects.*

Windows Bitmap Exports

Windows bitmaps (BMP files) are a totally raster file format exported from the layout. All points, lines, and polygons are converted into a bitmap, at the resolution you specify under the Options feature in the Export dialog box. BMP files can be inserted in Microsoft Office documents (but not edited as easily as WMFs) and opened by Corel Draw or Hijack Graphics.

PostScript File Exports

Three types of PostScript file exports are offered by ArcView 3.2, and they work in a variety of software programs. PostScript files, being resolution independent, are the export of choice for use in other graphic design programs, where graphics and text can be edited, and colors adjusted.

You should know the distinction between opening a PostScript file and placing a PostScript file. Usually, you will want to open a Post-Script file in Freehand or Illustrator by going to the File menu in either program and selecting Open. The file should open and become editable in the program. If you do the same in a word processing program, the PostScript file will not come up as an editable graphic, but as a lengthy list of PostScript code. This is because PostScript is an ASCII-based programming language, and the actual code is editable in a word processor.

To include or embed a PostScript graphic in a word processing document, use the Place or Insert option. The graphic should normally become visible as a graphic. With ArcView EPS exports, a grayed-out box with a large X or a message will appear. The message explains: "This EPS Picture was saved without a preview in it." The graphic will print correctly on PostScript printers, and can be sized or scaled on the page, but cannot be edited or seen on the page. The PostScript graphic needs to be opened and saved again in a graphic design program in order to have a preview associated with it.

The PostScript New (EPS) option features Postscript Level II file exports, with options for 300, 360, 600, and 720 dpi resolution. The resolution settings should not affect the resolution of any of the vector graphics in the map, but apply to the resolution of the raster graphics such as aerial photos or digital elevation models.

The basic PostScript (EPS) option results in a PostScript Level I file, and does not have the resolution options of the PostScript New option. One reason to use this option is that ArcView can take PostScript Level I files and place them back in a layout using the Picture Frame tool in the layout document. The Adobe Illustrator export option produces a PostScript file, called an *.ai* file, which is usable in some versions of Adobe Illustrator.

ESRI Versus Adobe: Making Sense of PostScript Files

One of the long-suffering problems with ArcView has been its PostScript exports to other graphic design programs. Although the PostScript file format is an industry standard, software companies can and do add or subtract code from some PostScript files that make them unusable or erratic in other PostScript-based programs.

For years, the Adobe Illustrator export files from both ArcInfo and ArcView were not openable in Adobe Illustrator. Today, Adobe Illustrator 8.01 for Windows can open some of these file exports from ArcView 3.2, but colors will change and text can become uneditable.

You will have mixed success opening both PostScript New and basic PostScript file exports from ArcView with Adobe and MacroMedia software products. Text objects are frequently converted into graphics, making them thicker and darker than normal fonts, as well as being uneditable. When text is converted as text, it may appear as individual, unconnected letters rather than entire words or sentences, making changes difficult or impossible.

ArcView PostScript files can be opened normally in Corel Draw, a graphics program not known for producing superior PostScript of its own. If you really need high-quality PostScript output from ArcView, your options are to perform several work-arounds with the export files or purchase Adobe Acrobat Writer, which allows ArcView (and other programs) to print directly to an Adobe PDF format.

Computer Graphic Metafile Exports

CGM (computer graphic metafiles) are a cross-platform graphics standard created in the early 1990s. ArcView produces three CGM export types: CGM Binary, CGM Character, and CGM Clear Text. Image resolution can be set in the Export options from 300 to 720 dpi. Although still useful for transferring graphic files between software programs on UNIX platforms, or from UNIX boxes to PCs, this file type has been largely supplanted by Windows metafiles in the PC world.

JPEG Exports

One of the most useful export formats is the JPEG (Joint Photographic Experts Group) format. This export option, shown in the following illustration, converts everything in the ArcView layout, both raster and vector, into a raster JPEG format, which is very popular on the Internet. JPEG files can be edited in Microsoft PhotoEditor, Adobe Photoshop, and other image manipulation programs. Because everything is converted into a raster format, you will lose editing control over lines and text.

JPEG Export feature and Image Quality options.

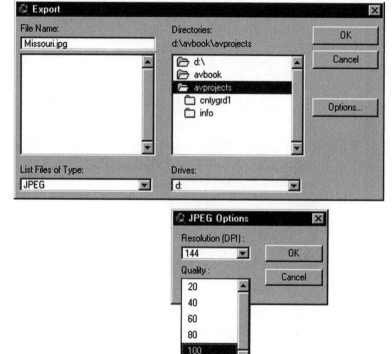

JPEG format is a highly compressed raster format that is not the best option for graphics work. That is, the higher the file compression, the more data the JPEG algorithms will discard (lose) from the original graphic. The ArcView JPEG export options allow you to set a final resolution of 72, 96, 120, or 144 dpi and to choose an image quality within that resolution setting. Image Quality ranges from low to high, including 20, 40, 60, 80, and 100% quality. With the 100% quality, no data will be discarded from the original graphic, whereas with 20% quality a lot of data will be thrown out.

Depending on the amount of detail in an ArcView map, final JPEG file sizes can vary from 100 kilobytes to several megabytes. Remember that JPEG files are compressed! A 2- or 3-megabyte JPEG file, when opened in a web browser or a graphic program such as Photoshop, can explode to 20 or 30 megabytes or more.

In general, ArcView maps with large areas of the same color will compress the best with little loss of detail, whereas finely detailed maps, including satellite images and color photos, will compress the least before losing important details. Keep in mind what you need the final JPEG file for (web pages, CD-ROM applications, and so on) and experiment with different levels of quality and file resolutions.

PICT Exports

Macintosh Picture format (PICT file) exports are only available on the old version of ArcView (ArcView 3.0a) found on the Macintosh operating system. PICT files are raster files similar to JPEG files, but without the compression options. PICT files can be edited by many photo-editing programs or placed into word processing documents on the Macintosh.

Portable Document Format Maps

There is yet another means of getting layouts out of ArcView and making them more usable. Acrobat 4.0, selling for about $200 from Adobe Systems, will let you "print to PDF" directly from an ArcView layout window.

There are several major advantages of having this capability. Adobe's portable document format or PDF files give you true cross-platform file viewing. A map printed to a PDF and given to another person on a CD-ROM, Zip disk, or floppy disk can be opened, viewed, and printed using Adobe Acrobat Reader, a PDF viewing software program freely downloadable from the Adobe web site.

Using Adobe, colors stay close to their original hues in ArcView, the map has the proper page size, and text stays in the right places, even if the other computer does not have the same fonts installed, or is running a totally different operating system. PDF maps made in ArcView can be saved as archives of past projects, grouped into larger PDF documents, and retrieved and reprinted as needed.

ArcView maps printed to "screen-resolution" level PDFs can be loaded into a web page, and viewed directly by Acrobat Reader from within another person's web browser. Higher-quality PDFs saved as print resolution files can be downloaded from a web page and printed clearly on most office printers. If map publishing is the goal, press-ready PDFs can be generated from ArcView, which saves information needed for 1,200- or 2,400-dpi high-resolution image setters.

The PDFwriter Process

Once it is installed correctly on your system, Acrobat PDFwriter works like any other printing device, but instead of sending the layout to a physical printer, PDFwriter writes the layout to a PDF file on your hard drive. After finishing your ArcView layout, you access the PDFwriter driver from the Print Setup dialog in the File menu. Acrobat PDFwriter should appear as an option in the Printer Name drop-down menu in the Print Setup dialog. Once the PDFwriter option is chosen, click on the Properties button to access the unique printing features. Four tabs (or windows) are available. The first window that appears is the PDFwriter Page Setup tab, the options of which are shown in the following illustration.

*PDFwriter Page
Setup options.*

Here you can choose a page size and orientation, and set margins for your PDF file. If you intend to eventually print the PDF resulting from this process on one of your office printers, remember to choose a page size and margin your printer can handle. Graphic resolution and scaling is set in the lower right of the dialog box.

Consider what you want to use the PDF file for. If you need the PDF for a simple web page, choose a lower graphic resolution such as the "screen" option (usually 72 dpi) or 150 dpi. If you intend to publish the map, choose a higher graphic resolution of 300 to 600 dpi. This choice of graphic resolution also influences what to do next in the Compression Options dialog box, shown in the following illustration.

*PDFwriter
Compression
Options
dialog.*

The Compression Options control how raster images embedded in the layout are handled. The most important choice here is the type and level of compression for the raster image, the choices being either Zip or JPEG. Zip compression is a "loss-less" method of compression, and has only one level. JPEG compression is more common, and has five levels of lousy compression, which means data will be thrown out at higher levels of compression. The JPEG options are High, Medium-High, Medium, Medium-Low, and Low.

If you want to place the resulting PDF on a web page used mainly for screen viewing, choose a higher level of image compression. The PDF image will load faster on the web page. If you need the PDF file for publishing, or for use in another graphics program, choose a lower level of image compression. More image data will be left to work with in the publishing process, or when the PDF file is opened in another program.

The third window is the Font Embedding tab, the options of which are shown in the following illustration. Fonts can either be embedded in a PDF or not embedded in the file. Not embedding fonts results in a smaller file size, and Acrobat will attempt to substitute fonts when the file is opened on a computer without the original fonts present. With ArcView maps, you should always embed fonts in the PDF, because correct feature labeling is dependent on precise positioning of text labels, and many symbol fonts (such as highway shields) cannot be replaced by substitute fonts.

PDFwriter Font Embedding options.

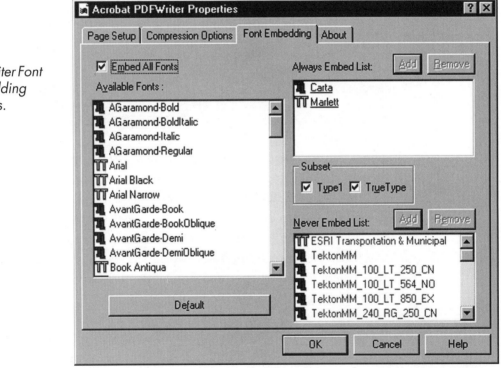

Embedding fonts in the PDF file also allows you to search the text on a map using the Find option in Adobe Acrobat Reader, a powerful feature.

PDF Exports to Graphic Design Programs

Adobe Acrobat PDFwriter also solves the irksome problem of inconsistent PostScript exports from the ArcView layout. PDFwriter uses a software subsystem called Distiller that acts as a PostScript software RIP. This subsystem cleans up or distills the PostScript code from ArcView and results in a PDF document that can be easily read in Adobe Illustrator or Adobe Photoshop.

Printing the ArcView layout to PDF can even handle large raster images and convert them to easily used files for other graphic programs. This process, shown in the following illustration, is discussed in the material that follows.

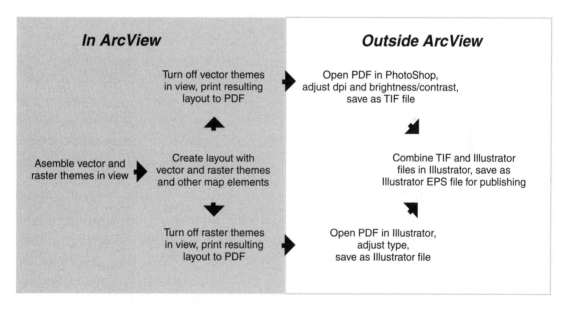

Process for exporting images in layouts with Acrobat PDFwriter.

You have set up a finished ArcView layout with several vector data themes overlying a large raster theme, such as an aerial photograph or a digital elevation model. You need to work with this file in a Post-Script graphic design program to get it ready for a publishing in a book. Export the raster image first. Return to the view document and turn off all vector themes, but keep the raster image visible. In the layout, print the layout to PDF, and open the resulting PDF file in Adobe Photoshop. Resave the file in Photoshop as a TIF file.

Return to the ArcView view document and turn on all vector themes and turn off the raster image. In the layout, print the vector themes to PDF and open the resulting file in Adobe Illustrator. While in Illustrator, open the raster image saved as a TIF and copy it into the new Illustrator file that contains the vector data. Now you have complete control over both the raster and vector data in a high-end PostScript graphic design program, and you are ready to send the Illustrator file to a publisher.

Stand-alone Hyperlinked PDF Map Collections

You can hyperlink maps printed to PDF files together in one large collection of PDF documents, and save this collection to a CD-ROM for distribution. Maps can be interspersed with text pages in a report saved as a PDF, with appropriate hyperlinks between text and maps. Hyperlinks on ArcView PDFed maps can be set in Acrobat based on a piece of text on the map, or on an area of the map selected as a "hot button."

The major advantage of producing PDF map collections is the high resolution of the PDF products. Very detailed maps using digital raster images of topographic maps or aerial photos can be distributed on CD-ROM throughout a business or to potential customers, without the end user needing additional software, for viewing or printing, other than the free Adobe Acrobat Reader. This is a great way of making complex ArcView map products widely available, without compromising map quality.

Maps for the Web

Maps are becoming a major attraction for many sites on the World Wide Web, and ArcView gives you several options for producing high-quality web products. Web maps can either be static, dynamic image maps, or interactive maps. The sections that follow discuss these three types of web map.

Static Web Maps

Static maps are simple raster images of ArcView maps, placed on a web page for viewing or downloading. The easiest method of preparing a static map for the web is by using the JPEG export in the ArcView layout. Depending on the complexity of the map and the page size of the image, the amount of image compression can be adjusted in the Export dialog to make the map appear in the web browser more quickly.

For even simpler maps, convert the JPEG map image to an indexed color GIF file (graphics interchange format) in any popular paint program, such as Microsoft Photo Editor, Adobe Photoshop, or Paint Shop Pro. Indexed color GIFs are small in file size compared to other image types, and can have transparent (rather than white) backgrounds. Images can be saved as interlaced GIFs that appear as low-resolution images in a web browser, while the full image downloads onto the page.

As mentioned previously, ArcView maps printed to PDF files are great for providing downloadable high-quality map products. PDFs embedded in web pages can work as static map images as well, but require the user to have the Adobe Acrobat Reader plug-in for their web browser on their computer. Any static map placed as an image on a web page can be downloaded using the "Save Image as" feature in the File menu of your web browser.

Imagemaps

One step up from static web maps are imagemaps. Imagemaps are graphics that contain several hotspots that when clicked on lead the user to another window of information on another web page. Imagemaps operate almost exactly like hotlinks in ArcView, except that they are designed to work on the web with URL addresses. The hotspots are normally polygon features on a map, but can be associated with point features as well.

There are several design considerations for web imagemaps to keep in mind. Keep the map simple in detail, like a very basic thematic map, and not complex or cluttered. Make the map features that will function as hotspots obvious graphically, using good figure-to-ground contrast, for example. Alternatively, use explanatory text to prompt the user to click on a hotspot, such as "Click on the cities on the map for more travel information." A web browser imagemap is shown in the facing illustration.

Generating imagemaps is not a built-in feature in ArcView, but there are several Avenue scripts available on the ESRI ArcScripts web page that do a good job. One of the better ones is *poly2imap.ave*. Once downloaded to your hard drive, you can open a new Scripts document in ArcView, load the text file *poly2imap.ave*, and compile the script.

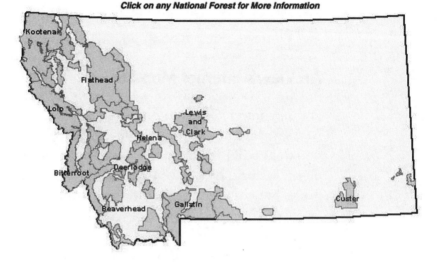

A simple imagemap appearing in a web browser.

Unlike most maps, imagemaps are produced in the view document and not in an ArcView layout. In the view, choose the polygon theme you want to convert into the imagemap with hotspots. Go to the theme's attribute table and edit the table to add a string field (a text field) that will contain the URL to the web pages the hotspots will lead to. Those web pages should already exist, either created by you for your own web site or as part of another web site on the Internet.

Save the table edits and run the complied *poly2imap.ave* script. The script's dialogs will prompt you to save the exported imagemap as a JPEG file, and ask you to name the accompanying *html* web page. Last, the script's dialog asks you to identify the tag field that holds the URL links.

After the script has completed its run, you are left with two files created outside ArcView; an *html* document with an embedded JPEG imagemap, and the image itself as an individual JPEG file. Open the *html* document in your web browser and the hotspots on the imagemap should already be linked to other web pages. This *html* document can be edited to add text or other graphics to complete the web page.

ArcView's Internet Map Server

For fully interactive maps on the web, ArcView users can turn to the ArcView Internet Map Server or AVIMS. ArcView IMS is an extension from ESRI that serves ArcView view documents, but not layouts, on the web. As a result, you are not serving finished maps on the web per se, but changeable collections of themes the end user can turn off or on, and query interactively.

ArcView IMS is one of several Internet mapping solutions available from ESRI. Unlike other Internet mapping solutions, AVIMS can be set up on a web server very quickly, literally in an hour or less, and requires no up-front programming. If you know how to set up an easily understandable view document in ArcView, you can serve it on the web.

The disadvantages are that AVIMS can be considerably slower than other mapping servers when dozens of users are accessing it at the same time, and you have fewer options for customizing the AVIMS interface than with software such as ESRI's MapObjects. All things considered, ArcView IMS is a quick, cost-effective alternative for serving interactive maps on the web. An example of this is shown in the following illustration.

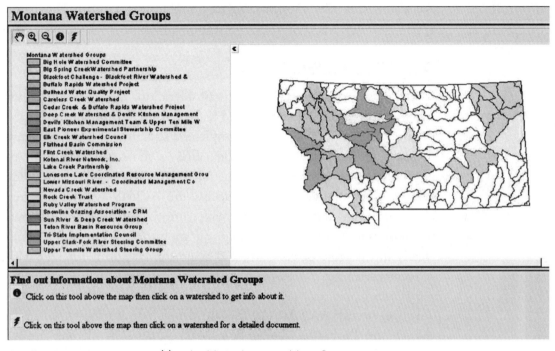

An ArcView project served by ArcView Internet Map Server.

AVIMS Tools

The standard ArcView IMS interface consists of a view window with a table of contents and map display. Themes in the table of contents can be checked on or off (made visible or invisible) and reshuffled in drawing order, but the user on the Internet does not have access to the Legend Editor. Therefore, it is up to you as the cartographer to symbolize the themes in a meaningful way that is easily understandable by the Internet user.

The user has the choice of the basic ArcView navigation tools: zoom in/out with the magnifying glass, and the pan or hand tool. Users can use the Identify tool to query map features in themes the designer has allowed to be queried. Simple letter-size maps can be printed from the Print tool. The Find and Query tools use a custom Java-based applet to search selected themes' attribute tables for information.

Design Considerations

There are several important design considerations to take into account when constructing AVIMS maps. First, keep the symbolization of the themes used in the map display simple and straightforward. Make sure to use a web-safe palette of colors, such as the *safety.avp* palette, to ensure that map features maintain the same appearance on the wide variety of computers that will access the AVIMS web site. When choosing point marker symbols, make sure to choose larger symbols that will show up on low-resolution monitors. Name the themes clearly in the table of contents, and avoid the cryptic file names with *.shp* extensions.

Take full advantage of ArcView's scale-dependent display properties in the Theme Properties dialog box. Make small-scale themes appear first when a user enters the AVIMS web page, and allow more detailed, large-scale data to show up after people zoom into a smaller portion of the map area.

Finally, always check out your AVIMS web site from several different computers with different monitors, operating systems, and Internet connections to see if it is showing what you intend to show. See plate 16 of the color insert for several examples of AVIMS maps.

Summary

Mapmakers are no longer limited to simple black-and-white, letter-size pages produced on an office printer. Cartographers using Arc-View now have the choice of producing dazzling full-color paper maps from letter to poster size, high-quality published map products, and web-based interactive maps. One primary design consideration remains. Know your map audience and use the modern digital tools in ArcView to meet the needs of your map users.

Index

A

A0 size 223
A1 size 223
A2 size 223
A3 size 223
A4 size 223
Acrobat PDFwriter 339
Add Class button 58
Add Theme button 34, 46, 286
    using 146
Add Theme dialog box 46
adding data as themes 34
Adobe Acrobat 338
Adobe Acrobat Reader 309, 339, 345
Adobe Illustrator 122, 162, 165, 333, 335
    export files 333, 335
    PDF files in 343
Adobe Photoshop 292, 336, 345
    PDF files in 343, 344
Adobe PostScript page description language 307
Adobe Type Manager (ATM) 125–127
ADRG files 286
aerial photography 280, 296
AI files 333, 335
aiapress.avx 312
AIshd to Avcolor 111
Albers Equal Area projection 172, 194, 199
altered projection types 174
amfm.avp 95
angled type, guidelines for 133
annotation themes, from ArcInfo and AutoCAD 146–
        147

ArcInfo
    annotation themes from 146–147
    GRID files 282
    naming shapefiles in 49
ArcPress 307, 311–317
    accessing 312
    color printing 316–317
    Page Layout options 315
    Print dialog box 314
    printer file types supported 312
    printing from 322
    printing quirks 317
    raster file export options 314
    temp files in 313
ArcScripts page, ESRI web site 102, 111, 119, 185, 261,
        262, 312, 346
ArcView Advanced Label extension 2.3 (AVALabel
        2.3) 142, 150, 162–164
ArcView Internet Map Server (AVIMS) 348–350
    design considerations 350
    tools 349
ArcView ODB extension 228, 267
    See also Save and Restore extension
ArcView Projection Utility extension 186, 188, 189,
        200–210
    accessing 201
ArcView Spatial Analyst 21, 287
    used to generate contour lines 51
arrows.avp 95
Artist palette 111
artist.avp 252, 254, 330
Attach Graphics option, using 151

attribute information, in metadata 14
attribute tables
    placing records from (Mapper) 271
    used for labeling 148–153
AutoCAD, annotation themes from 146
Auto-label dialog box 155
Auto-label option 154–158
    caution about overlapping text 157
    examples of 156
    used with line themes 156
    using 141
AVALabel 2.3. *See* ArcView Advanced Label extension
Avcolor 111
Avenue scripts
    for generating imagemaps 346
    for storing tabular data as decimal degrees 177
AVIMS. *See* ArcView Internet Map Server extension
AVP files 96
azimuthal projections 171
    illustrated 173
    Lambert Equal Area 173

B

Background color 100
Banner Label icon 142
Banner Text icon 142
Banner Text tool 145
bar charts 72
base maps
    legend type for 55
    using DRGs as 293, 295
Behmann projection 193
Bevel option (Pen) 114
BIL files 237, 286, 314
BIP files 237, 314
Birmy Graphics PowerRIP 311, 318, 319
bit depth, defined 283
Blue Marble Geographics extensions 211
BMP files 237, 286, 314, 333, 334
brightness, in raster images 289, 295
Bring to Front 239
BSQ files 237, 286, 314
buffers
    for cartographic effects 249
    for land features 253–255
    for water 250–253
Bullet Leader icon 142
Bullet Leader Text icon 142
Bullet Leader Text tool 145
Butt option (Pen) 114

C

C size 223
c256.avp 95
CADRG files 286
Calcomp Electrostatic 312
Calcomp Techjets 312
calibrat.apr 104, 330
calibrating color 104
calibration, printer 330
California Department of Fish and Game 294
callout box, default style for 144
Callout label icon 142
Callout Text 142, 144
camera size 223
Canon Bubblejets 312
Canon Inkjet printers, software RIP for 318
Cap option 114
carto.avp 94, 108, 295
    for PostScript fills 324
cartographic design process 32–38
    checklist 34
cartography
    ArcView tools for 15
    brief history of 2–5
Cassini Soldner projection 199
CBJ files 312
CCRF files 312
CCRF-IL files 312
cells, in raster data 20
Central Meridian (View Properties) 178
CGM files 333, 336
Chart Frame tool 234–235
chart maps 52
chart symbol legend type 54, 72–73
    limitations 73
Chasen, Robert 248
choropleth maps 50
CIB files 286
Clarke's 1886 spheroid 175, 194, 196
class symbols, sizes 64
classes, number of 78
classification 75–87
Classification dialog box 78
Classification Field drop-down list 57
classification methods
    choosing 36
    compared 79
    equal area 84–86
    equal interval 82–84
    natural breaks 79–81
    quantile 81
    standard deviation 86–87

types of 75, 78–87
Classify button 62
Clip to View 228, 229
ClrRamp0.avx 248
CMYK color model 24, 40, 100, 101, 103
color
    calibrating monitor for 104
    complexity of 92
    customizing in legends 260
    for land buffers 254
    for publication printing 40
    for water buffers 252
    in annotation themes 146
    in PDF files 339
    perception of 141
    printed from ArcPress 316
    smoothing gradations of 259
    standards in mapmaking 24, 99
    use of 97–106
    variability with software RIPs 311
color images 285
color inkjet printers 325
color laser printers 325
color maps, altering 291
color matching 62, 105
color models 100–104
    compared 103
    defined 24
Color palette 97–106
    chart symbol legend 73
    custom colors 98
    default colors 98
    fill options 98
    specialized 95
Color Picker extension 101, 102–104
color plotters 327
color printing 302
color ramps
    customized 62
    predefined 59–62
    viewing 292
color schemes, predefined 67
color separations 329
color values
    contrasting 41
    on paper maps 41
    on transmissive media 41
colormrk.avp 95
colornam.avp 95
ColorRamp extension 248, 259, 260–261
colors, isolating (DRG-Tools) 295

compasses 233
composition of map elements 215–220
compression
    for the Web 345
    option with PDFwriter 341
Computer Graphics Metafiles 333
    exporting to 336
conic projections 171
    Albers Equal Area 173
    illustrated 172, 173
continuous distribution of data 77
contour lines, generating 51
contour maps 51
contrast, in raster images 295
Convert to Shapefile command 192
    to copy original data 47
coordinate systems 176–177
    geographic 176
    global coordinate system 204
    projected 176, 177
    single zone vs. multiple zone 196
    using ArcView Projection Utility 203
Copy Themes command 133
copying
    data, precautions for 47
    shapefiles 133
Corel Draw 333, 336
COS.Legend-Smooth-Continuous 248
Create Buffers command 51, 250
Create Buffers dialog box 251
Create Markers tool 110
Create Overview 276
createcolors.avx 102
crime.avp 95
curved type, guidelines for 133, 146
Custom 262, 267
Custom button (Color palette) 100
Custom Legend Tool extension 262–266
cylindrical projections 171, 172
    illustrated 172

D

D size 223
DAK. *See* ESRI Data Acquisition Kit
darkness, in raster images 289
data
    adding as themes 34, 46
    copying 47
    decimal degree 187
    documentation of 13
    inadvertent corruption of 47

normalizing 50, 57, 58
   obtaining 33
   projected 186, 205
   unprojected 186, 187
data distribution 75, 76
data layers, listing 33
data quality information, in metadata 14
Data Source Type (Add Theme) 286
data sources 33
data types
   comparison of 20
   raster 20
   vector 18
Datum tab (Projection Utility) 206
datums, geodetic 176
   changing 176
   local 176
decimal degrees 177, 184, 191
   for flexibility in changing projections 187
defaults
   callout box style 144
   classification method 78
   fills 97, 108
   in Auto-label dialog box 157
   Legend Editor classes 58, 63
   legend type 55
   line ends 113
   line types 113
   line weights in Custom Legend Tool 265
   map titles 161
   number of legend columns 264
   output resolution 305
   palettes 96, 97
   point markers 109
   polygon outlines 98
   printer drivers 307
   quantile classification 81
   text and labels 154
   text style in TOC 140
   Text tool 143
   type size in Custom Legend Tool 264
Delaune, Mike 248, 250
Delete Class button 59
DEM. *See* digital elevation models
design elements 215
design enhancement techniques 248–276
   dissolving features 255–257
   generalizing themes 257–259
   using buffers 250–255
   using Legend tools 259–261
   using multiple copies of a theme 248, 249
design principles 26–32

figure-ground contrast 28
   hierarchical organization 29, 110, 113
   legibility 26, 94, 113
   simplicity 92
   visual contrast 27
   visual hierarchy 93
design process. *See* cartographic design process
design tools 22–26
dichromatic color ramps 60
digital elevation models (DEM) 282, 283
   defined 21
digital imagesetter. *See* imagesetter
Digital Ortho Quarter Quads (DOQQs) 297
digital orthoquad maps (DOQ) 280, 296
digital raster graphics (DRGs) 282, 293–296
   features and colors 295
   illustrated 294
   printing 296
direction (rotated) of symbols 64
Display option (Legend Frame) 230
display, refreshing 229
dissolves
   for cartographic effects 255–257
   results of 257
   uses of 256
distance units, setting 184, 228
Distiller 343
distribution of numerical data 75, 76
distribution information, in metadata 14
documentation tools 15
DOQQs. *See* Digital Ortho Quarter Quads
dot density legend type 54, 68–71
dot density maps 51, 68–71
Douglas-Peucker algorithm 257
Draft option (View Frame) 229
DRG. *See* digital raster graphics (DRGs)
DRG-Tools extension 294
Drop Shadow Text tool 142, 145

E

E size 223
edgematching (DRG-Tools) 294
electronic media 13
elements, map 216
   aligning 225
   in map composition 216
   positioning of 218, 219, 239
   using neatlines with 220
elevation, color ramp for 61
Ellipsoid tab (Projection Utility) 208
Enable Clipping 333
Encad Novajets 312

Enhanced PostScript Level II 307, 321
envtl.avp 95
EPS file statistics 313
EPS files 39, 237, 313
Epson Stylus Pro Inkjets 312
Epson Stylus RIP 309, 311, 319
equal area classification 75, 84–86
equal area cylindrical projection 199
equal area projection 179, 193
equal interval classification 75, 77, 82–84
  scheme for 64
Equal Spacing (Graphics) 240
equatorial projection 195
equidistant azimuthal projections 193, 199
equidistant conic projections 194, 199
equidistant cylindrical projections 199
equidistant projection 180
ERDAS 21
ERDAS file types 286
ESC/P2 files 312
ESRI Data Acquisition Kit (DAK) 210
ESRI web site 102, 111, 185, 262
even distribution of data 77
Export (File) 332
Export dialog box 332, 334
exporting
  files for printing 331–338
  files with PDFwriter 343
  PostScript files 334
extensional organization 30, 31
extensions
  ArcPress 311–317
  ArcView Advanced Label 2.3 (AVALabel 2.3) 142,
    150, 162–164
  ArcView Internet Map Server (AVIMS) 43, 348–350
  ArcView ODB 228
  ArcView Projection Utility 186, 188, 189, 200–210
  ArcView Spatial Analyst 287
  Color Picker 101, 102–104
  ColorRamp 259, 260–261
  Custom Legend Tool 262–266
  DRG-Tools 294
  for use in layout (list) 267
  for use with ArcView (list) 248
  Generalize 258
  Geoprocessing 255
  Graticules and Grids 267–270
  Image Analyst 21
  Legend-Smooth-Continuous 259
  Mapper 270–272
  Multi-theme Auto-label 158–161
  Named Extents 185

Nudger 274–275
Overview 275–276
Save and Restore 272–274
ShapeProjector AVX 211
Spatial Analyst 21, 51
Symbolizer 116–118
Transformer AVX 211
Xtools 250

F

false easting (View Properties) 179
false northing (View Properties) 179, 205
feature labeling, defined 123
Federal Geographic Data Committee Standard
    (FGDC) 13–15
FGDS. *See* Federal Geographic Data Committee
    Standard
Field Calculator, using 152
figure-ground contrast 28, 37, 99
file compression for the Web 345
file names
  8.3-character limitation 48
  conventions for 49
file types
  ADRG 286
  AI 333, 335
  AVP files 96
  BIL 286, 314
  BIP 314
  BMP 286, 314, 333
  BSQ 286, 314
  CADRG 286
  CBJ 312
  CCRF 312
  CCRF-IL 312
  CGM 333, 336
  CIB 286
  DBF files 48
  EPS 333
  ESC/P2 312
  geotiff 298
  GIF 345
  GIS 286
  GRID 282, 286
  IMG 286
  JPEG 287, 314, 333, 336, 337
  LAN 286
  LZW 287
  MrSID 287
  NITF 287
  PBM 314
  PCL3-PCL5 312

PCX 314
PDF 309, 321, 336
PGM 314
PICT 333, 338
PNG 314
PPM 314
PRJ 189, 202
PRT 321
RAS 287
raster files for export 333
RGI 312
RIC 312
RLS 286
RS 287
RTL 312
SHP files 48
SHX files 48
SUN 287
TIF 287, 314
vector files for export 333
VRF 312
WMF 333
files, large, problems with 255
Fill palette 97–98, 107–109
    specialized 94
fill patterns 67
    customized 117
    printing 324
    setting properties 143
Fill View Frame 228, 229
Flip Symbols button 59
Font Embedding tab, with PDF files 342
font families 124
font names 124
Font palette, default 27
fonts
    bitmapped 124
    defined 123
    PostScript 125
    Truetype 125
Foreground color 100
    changing 110
forestry.avp 94, 95
formatted text, copying into ArcView 162
Frame Properties dialog box 226
Frame tool, using 38
Frame tools 225–238
    Legend Frame 226
    Scale Frame 226
    View Frame 226

frames, defined 213
full spectrum color ramp 62

G
Generalize dialog box 258
Generalize extension 248, 258
generalize.avx 248
generalizing themes 257
    results 259
    uses of 258
generic square/oval highway labels icons 142
geodetic datums 175, 176
Geodetic Reference System 1980. *See* GRS80
geographic coordinate systems 176, 191
    with ArcView Projection Utility 204
geographic data layers. *See* data layers
geographic features
    interference from labels 131
    labeling 21, 122
    size manipulation of symbols 19
    symbolization of 7, 18
Geology palette 115
geology.avp 95
geoproc.avx 248
Geoprocessing extension, dissolve operation 255
Geoprocessing Wizard 248
geotiff files 298
GIF files 237, 345
    indexed color 345
GIS files 237, 286
Global Coordinate System 204
global positioning system (GPS) 176
    using data from 211
Gnomic projections 193, 199
government web sites 15
GPS. *See* global positioning system
graduated color legend type 54, 57–62
graduated symbol legend type 52, 62–65
    advanced options 64
    customized ramps 64
graphics
    file types supported 237
    PostScript 307, 334
    resizing 226
    uses of 238, 243
graphics tools 238–244
    Align 239
    Bring to Front/Send to Back 239
    drawing 242
    Group/Ungroup 241

neatlines 241
Simplify 241
size and position 238
gratgrid.avx 267
Graticule and Grids extension
limitations of 267
graticules
defined 177, 267
illustrated 270
setting parameters 269
Graticules and Grids extension 267, 267–270
Graticules and Measured Grids extension
See Graticules and Grids extension
Gray button 292
grayscale images 284
grayscale laser printers 323
grayscale printing 302
grayscale ramps, setting up 292
grayscale values, contrasting 41
grid. *See* measured grid
GRID files 210, 282, 286
working with in ArcView 287
GRS80 175, 194
guidelines for aligning elements 240

H

Hammer Aitoff projection 193, 199
Handmade Software 320
hardware RIPs 309–310
hardware.avp 95
hatch fills, using 249
hatch.avp 94, 108
hazmat.avp 95
Hewlett-Packard Designjet plotters 327
Hewlett-Packard Designjets 312
Hewlett-Packard Deskjets 312
Hewlett-Packard Laserjets 312
Hide/Show Legend option (Theme) 139
hierarchical organization 29, 110
highway symbols 153
Hijack Graphics 333
Horizontal Alignment (Text Properties) 144
hot button, setting map area as 344
hot spots, on imagemaps 346
Hotine Oblique Mercator projection 199
HP 1050CM Designjet plotter 328
HSV color model 24, 40, 100, 103
Huber, Bill 248
hue variables
defined 24
illustrated 23

hue, defined 100
hydrographic buffers. *See* water buffers
hyperlinked maps 344

I

icons.avp 95
identification information, in metadata 14
Identity tool 290
Image Alchemy PS 320
Image Analyst 21
Image Colormap (Image Legend Editor) 291–292
image data types 283–285
grayscale 284
monochrome 284
multi-band 285
pseudocolor 284
true color 285
used in ArcView (list) 285
image file formats
types of 46
types supported in ArcView 286
Image Legend Editor 287–292
Identify tool 290
Image Colormap 291
Interval Lookup tool 290
Linear Lookup tool 288
using 21
image resolution. *See* Page Setup dialog box,
output resolution
imagemaps 43, 346–348
generating 346
illustrated 347
imagery
multi-spectral 280
scanned 281
images, adding in ArcView 286
imagesetters 39, 329
IMAGINE files 286
IMG files 286
IMPELL files 286
indexed color 345
indexed color images 284
inkjet plotters 327
capabilities and limitations 328
ink 329
inkjet printers, color 325
interactive maps 348
interactive media 13, 42
Internet, maps for 345–350
Internet Map Server Extension (AVIMS) 43
Internet sources

for ArcPress 312
for Avenue scripts 119
for creating symbols 116
for DOQQs 298
for extensions 102
for extra palettes 111
for generating imagemaps 346
for layout enhancements 262
for low-cost images 293
for metadata information 15
for setting reference scales 185
Interval Lookup tool 290
Iris prints 310
isarithmic maps 51
isolines, defined 51

J

Jenks Optimization algorithm 79
Join option (Pen) 114
Joint Photographic Experts Group format.
    See JPEG files
JPEG compression 341
JPEG files 287, 314, 333
    exporting to 336
    for the Web 345
    options 337

K

kerning 135

L

Label tool 141
    drop-down palette 142
Label tools 148
    highway symbols 153
labeling map features 21, 37, 121–166
    advanced labeling 162–164
    ArcView methods for 141–158
    hydrographic 130, 131
    in Legend Editor 138
    process in ArcView 136, 141–158
    scale of 149
    using attribute tables 148
    using MaPublisher 165
    with Auto-label option 154–158
    with AVALabel 2.3 162–164
    with Multi-theme Auto-label extension 158–161
    with subsets 155
labeling margin information 161
labeling systems 128, 129
Labels column (Legend Editor) 59
Lambert Conformal projection 172, 180, 194, 196, 199
Lambert Equal Area Azimuthal projections 193

Lambert Equal Area projection 172, 199
LAN files 237, 286
land buffers 253–255
land cover, color ramp for 61
Landsat satellite image 280
large-scale maps
    changing scale in 188
    data in 184, 186
    example 183
    UTM projections for 197
laser printers
    color 325
    grayscale 323
latitude/longitude lines 134, 177
layers, isolating (in DRG-Tools) 295
layout document 213–246
    advanced techniques 262–277
    using 3
layout properties 222–225
Layout Properties dialog box 224
layouts
    naming 224
    saving as Windows Metafiles 333
    sketching 37
ledger size 303
legal size 223, 303
legend bar, smooth continuous 260
legend box, establishing parameters for 265
legend design process 263–266
Legend Editor 45–73
    accessing 52
    chart symbol legend type 72–73
    classification in 75–87
    Density Field 68
    dot legend type 68
    graduated color legend type 57
    graduated symbol legend type 62
    Labels column 59
    palettes 89–119
    Properties button 73
    Symbol box 64
    unique value legend type 65
    using 19, 36, 154
    using to control visual variables 22
    Value column 59
    Values Field 65
legend enhancement techniques 259–266
Legend Frame Properties dialog box 230
Legend Frame tool 141, 230–231
legend frames, resizing 230
Legend Tool extension 267
    See also Custom Legend Tool
Legend tools 259–261

Custom Legend Tool extension 262–266
for color 259
legends, map 8
clarity in 139
complex 66
creating text for 137
customizing color in 260
enhancing 231
in map composition 216, 218
style in 220
type in, choosing 36
types of 52–54, 73
legends.avx 262, 267
Legend-Smooth-Continuous extension 248, 259
legibility
designing for 26
in map composition 218
with line themes 113
letter size 223, 303
lettering, defined 123
line attribute tables, line segments in 157
Line Label Position option 157
line styles 140
line symbology 113–118
line themes
auto-labeling 156
offsetting a line in 65
using graduated symbols for 62
line weight, default 27
Linear Lookup tool 288–289
lines
curved, placing 145
defined 18, 19
illustrated 23
specialized 115, 117
lineset files 111
literal labeling 128
live links 227
live-linking to View document 227
locative labeling 128, 129
longitude-latitude lines
creating 267–270
LZW compressed files 287

M

Macromedia Freehand 122, 162, 165
Make Default button 97
map composition 215–221
balance in 218
elements in 216
for large vs. small-scale maps 216
legends in 218

neatlines in 220
North arrow 221
scales in 218
stored as templates 220
map display, in map composition 216
map elements, composition of 215–220
map extent properties 228, 229
map features, naming 145
MapObjects 348
map orientation 134
map products
in ArcView 12
presentation-quality 39
setting size 213
types of 11–12
map projections 9, 167–212
and scale 182–184
changing 180, 186, 187, 189
changing in View Properties 191
changing with batch process 201
concepts 170–179
considerations in ArcView 187–189
defined 168
determining 34
parameters 178–179, 198–200, 205
selecting 179–179
terminology 170–179
testing for accuracy 188
map projections, types of 171–175
Albers Equal Area 172, 194, 199
altered projection 174
azimuthal 171
Behmann 193
Cassini Soldner 199
conic 171, 172
custom 198–200
cylindrical 171, 172
equal area 179, 193
equal area cylindrical 199
equidistant 180
equidistant azimuthal 193, 199
equidistant conic 194, 199
equidistant cylindrical 199
Gnomic 193, 199
Hammer Aitoff 193, 199
Hotine Oblique Mercator 199
Lambert Conformal 172, 180, 194, 196, 199
Lambert Equal Area 199
Lambert Equal Area Azimuthal 193
list of 170
Mercator 168–169, 172, 193, 195, 199
Miller cylindrical 179, 193, 199

Mollweide 193, 199
national grids 194, 198
of hemispheres 193, 195
of the U.S. 194, 196
of the world 193, 195
orthographic 187, 199
Peters 193
planar equatorial 172
Plate Caree 193
pseudocylindrical 171, 174
Robinson 169, 174, 179, 182, 193, 195, 200
sinusoidal 193, 200
standard equatorial 195
State Plane 180, 196
State Plane 1927 194
stereographic 193, 200
Transverse Mercator 180, 196, 200
Universal Transverse Mercator (UTM) 194, 197
Vertical Near-Side Perspective 200
world-from-space 187, 193, 195
map purpose 11, 33
map series
extension for creating 273
locater map in 276
organizational style in 220
templates for 244
map servers 348
map units, setting 184, 228
MapObjects 189
Mapper extension 267, 270–272
Mapper zmapper.odb 267
See also Mapper Extension
maps
basic elements of 5–10, 216
finishing touches on 213–246
placing on page 226
types of 5, 43, 50, 345
MaPublisher 162, 165
margin information
adding 243
defined 161
labeling 161
typestyles in 220
margins
setting 223
size 303
Marker palette 109–110
specialized 95
markersets, creating 110
MCP files 237
measured grid
defined 267
setting parameters 269

media
electronic 13, 42
interactive 13
reflective 12, 41
transmissive 12, 41
types of 12
Mercator map projection 168–169, 172, 193, 195, 199
meridians, defined 177
metadata 13–15
and projection files 189–191
for DRGs 296
using 183, 184
using with ArcView Projection Utility 203
metadata reference information 14
Microsoft Office 334
Microsoft Photo Editor 336, 345
Microsoft PowerPoint 334
Miller Cylindrical projection 179, 193, 199
mineral.avp 95
Miter option (Pen) 114
Mollweide projection 193, 199
monochromatic color ramps 59
monochromatic images 284
MrSID files 287
multhlab.avx 158
multi-band images 285
multichromatic color ramps 60
multi-spectral imagery 280
Multi-theme Auto-label extension 158–161
municipl.avp 95

N

NAD_1983_UTM_Zone_12N 205
NAD_83_Montana 205
NAD27. *See* North American Datum of 1927
NAD83 176, 196, 204
NADCON. *See* North American Datum Conversion
Named Extents extension 185
namedext.avx 185
naming conventions 48–49
for geographic coordinate systems (Projection Utility) 204
national grids (map projections) 194, 198
Native OS printer driver 321
natural breaks classification 75, 77, 79–81
NBI files 237
neatlines 220, 241–242
settings for 242
NITF files 287
Nominal button 292
nominal labeling 128, 129
consistency in 130
normal distribution of data 77

normalizing data 50, 57, 58
North American Datum Conversion (NADCON) 176
North American Datum of 1927 176, 194, 196
North American Datum of 1983 194
North Arrow Manager 233–234
North arrows 10
  correcting direction of 234
  customizing 233
  in map composition 216, 221
north.avp 95
north.def 234
Nudger dialog box 275
Nudger extension 267, 274–275
null value symbols 56
numerical data distribution 75, 76

O

ODB Extension v1.2. *See* Save and Restore extension
ODB, defined 273
odb.avx 267
oilgas.avp 95
ordinal labeling 128, 130
organization of data layers 29–32
orientation variables
  defined 24
  illustrated 23
orthographic projection 187, 193, 199
outlines
  setting properties 143
  specialized 117
output, map 301–351
  for the Web 339, 345–350
  printing considerations 302–309, 334
  resolution 223, 335
  screen resolution 339
overhead transparencies, maps on 41
overlapping labels 157
overlapping type 129
Overview Utility extension 267, 275–276

P

page setup 222–225
  output resolution 223, 305
Page Setup dialog box 222
  output resolution 305
  setting margin size 304
page size 215
  changing 223
  determining 38
  printed 303
  setting 223–224
PaintShop Pro 345

Palette Manager 94–97
palettes 89–119
  accessing 91
  features of 91
  loading new files 96
  printing 118
  removing 96
  restoring defaults 96
  specialized 94, 111
Pantone Color Matching 319
paper for printing maps 330
paper maps 12
  color values on 41
parallels, defined 177
Parameters tab (Projection Utility) 206
patterns
  complex 92
  customized 117
  overprinting base map with 295
  progressions 108
  use of 107–109
PBM files 314
PCL3-PCL5 files 312
PCX files 314
PDF files 309, 321, 336
  for export 343
  for the Internet 42
  for web pages 345
  hyperlinked maps in 344
  printing to 338
  saved as TIF files 344
PDF map collections 344
PDFwriter 329, 339
  compression options 341
  Distiller 343
  Font Embedding options 342
PDFwriter Page Setup 339
  options 340
Pen palette 113–119
  default 27
  specialized 94
Peters projection 193
Petrotechnical Open Software Corporation 204
PGM files 314
PICT files 333
  exporting as 338
Picture Frame Properties box 237
Picture Frame tool 236–238
  graphics importable with 237
pie charts 72
pixel depth, defined 283
pixels, in raster data 20

Placeable WMF files 161, 333
planar equatorial projection 172
planar projection, illustration 173
planar surfaces 172
Plate Carree projection 193
plotters
   and spooling 322
   types of 327
PNG files 314
point markers, specialized 115
point symbology 109–110
point themes, using graduated symbols for 62
points
   defined 18, 19
   illustrated 23
poly2imap.ave 347
polygons
   defined 18, 19
   illustrated 23
   outlining 117
Portable Documents File format. *See* PDF files
POSC Code Number 204
positioning text 149
poster size 223, 303
PostScript
   defined 307
   Level I 307, 321, 333
   Level II 308, 321, 333, 335
   Level III 308, 321
   printer driver options 307, 308, 321
PostScript add-on board 310
PostScript Export option 39
PostScript File Exports 334
PostScript files, ESRI vs. Adobe 336
PostScript fonts 125
PostScript hatch patterns 295
PostScript interpreters, defined 309
PostScript New (EPS) files 333, 335
PostScript RIPs 309–310
   hardware 309–310
   hardware, external 310
   software 310, 343
PowerRip. *See* Birmy Graphics PowerRip
PPM files 314
precipitation, color ramp for 61
presentation quality maps 39
prime meridian 177
   in ArcView 195
Print dialog box (ArcPress) 314
Print Setup (File) 320

Print to File option 321
printer calibration 330
printer driver options 321
   PostScript 307, 308
printer speed 305
printer supplies, inkjet printers 327
printers, types of 323–330
   color inkjet 325
   color lasers 325
   color plotters 327
   grayscale lasers 323
   imagesetters 329
printing
   DRGs 296
   palettes 118
printing, map 301–351
   color vs. grayscale 302
   exporting files for 331–338
   from command prompt 321
   high-resolution color posters 310
   margins 303
   output resolution 305, 335
   page size 303
   paper for 330
   PostScript support 307
   process of 320
   spooling 322
   tips for 330–331
   with ArcPress 322
   with inkjets 322
PRJ files 189, 202
projected coordinate systems 177
   in ArcView Projection Utility 204
Projection (View Properties) 178
Projection button 191
projection coordinate systems 176
projection files 189–191, 202
   in Albers Equal Area, U.S. (list) 190
   in decimal degrees, U.S. (list) 190
   saving 207
projection parameters 178–179
projections. *See* map projections
Properties button 73
PRT 321
PS files 237
pseudocolor images 284
pseudocylindrical projections 171, 174
publication-quality maps 39, 330
   from PDFs 339
purpose, map 11, 33

Q

Quality option (Legend Frame) 230
Quality option (View Frame) 229
quantile classification 75, 77, 81
Quantitative Decisions web site 261

R

rainbow.avp 95
Ramp button 291
Ramp Colors button 59, 62
Ramp Symbols button 64
Random button 292
Random Symbols button 66
RAS files 287
raster capabilities in ArcView 21
raster data 20, 279–298
    compared with vector data 20
    types of 280–283
raster files
    BMP 334
    exporting from ArcPress 314
    GRID 282
    JPEG 338
    PICT 338
    supported for export 333
Raster Graphics 312
raster image processors (RIP) 307
    PostScript 309
raster images
    controlling brightness and contrast 288
    from outside sources 293–298
    on web pages 345
    saving separately 344
    showing relief in 290, 292
    statistics file in 288
    techniques for improving 292
    types of 284–285
raster.avp 94, 95
realty.avp 95
Reference Latitude (View Properties) 179
reference maps 5
    drafts 40
reference scales
    caution about 150
    example of setting 186
    selecting for Auto-label 155
    setting 64, 185
reflective media 12
    color on 41
refreshing display 229
reliefs, in raster images 290, 292
resizing graphics 226

resizing legend frames 230
resizing views 275
resolution, output 223
    and printing time 305
    high quality with PDFs 344
    with EPS 335
revising 38
RGB color model 24, 100, 101, 103
rgbtext.avp 95
RGI files 312
rhumb line 168
RIC files 312
RIP. *See* raster image processor
ripping a file 309
RLC files 237, 286
Robinson projection 169, 174, 179, 193, 195, 200
    illustrated 174
    limitations of 182
Rotation Angle (Text Properties) 144
Round option (Pen) 114
RS files 237, 287
RTL files 312

S

safety.avp 95, 350
Same Width-Same Height (Graphics) 240
sandwich technique 248, 249, 254
sans serif typestyles 123, 124
satellite imagery 280
saturation
    defined 101
    increasing 291
Save and Restore extension 267, 272–274
Scale Bar Frame Properties dialog 232
Scale Bar Frame tool 231–232
scale bars, custom (Mapper extension) 271, 272
Scale Frame Properties box 228
Scale Text with View (Text Properties) 144
Scale window 184, 185
scales, map 8, 180–186
    and projections 182–184
    considerations in ArcView 187–189
    defined 180
    determining 34
    in dot density maps 69
    in map composition 216
    large vs. small 183
    properties 229
    setting 184–186, 228
    table of equivalents 181
    using Mapper extension 271
scanned images 281

screen refresh time 229
sea floor elevation, color ramp for 61
Seagate Crystal Reports 236
secants 171
Send to Back 239
serif typestyles 123
shadeset files 111
shape variables 22, 23
shapefiles
    and projection information 189
    copying 133
    naming 48
    processing projections 202
Show 91
Show Symbol Window 91
Simplify tool 241
    used with legends 220, 231
single symbol legend type 54, 55
single-band images 285
single-color color ramps 59
single-color monochromatic color ramps 61
sinusoidal projection 193, 200
Size and Position (Graphics) 238
size range for symbols
    setting 63
    setting reference scale 64
size variables
    defined 23
    illustrated 23
size, chart 73
size, dot 69
size, line 113
size, map 12, 38
size, map files 122
size, marker 109
size, page 215, 223, 303
    changing 223
    list of common sizes 223
size, symbol, for the Internet 350
size, type 130, 155
    in annotation themes 146
    selecting 143
sketch, map layout 37
slides, maps on 12, 41
small-scale maps, data in 183
Smooth Gradations 259
Snap to Grid 225, 239
software RIPs 310–320
    ArcPress 311–317
    Birmy Graphics PowerRIP 319
    color shifts with 311
    Distiller 343

Epson Stylus RIP 319
    Image Alchemy PS 320
source statement 10
    adding 161
    in map composition 216
Spatial Analyst. *See* ArcView Spatial Analyst
spatial data, in metadata 14
specialized geographic color ramps 60, 61
Specify Color dialog box 99, 100
spheres/spheroids 174
    illustrated 175
Spheroid (View Properties) 178
Spline (Curved) Text icon 142
Spline Text tool 145
spreading type guidelines 135
Square option (Pen) 114
standard deviation classification 75, 77, 86–87
standard equatorial projection 195
Standard Parallels (View Properties) 179
standard parallels of latitude 172
standard, for metadata 13–15
State Plane coordinate system 177
State Plane projection 180, 194, 196
    illustrated 182
static maps 42, 345
statistics file, raster images 288
stereogrammic organization 29, 30
stereographic projection 193, 200
Sternberg, Howie 267
Store As Template feature (Layout) 224
style properties, TOC 140
subdivisional organization 31
Sun raster files 287
Symbol box, Legend Editor 64
Symbol dump script 330
symbol properties 140
Symbol Window 89
    accessing 91
    illustrated 90
Symbolizer extension 116–118
symbolizing data, basics of 18–21
symbols
    complexity 92
    customizing 110
    direction of 64
    in legends 8
    Internet sources for 111
    options for in Custom Legend Tool 265
    printing 118
    size of 63, 64, 69
    unique values of 65
symdump.ave 119, 330

T

Table Frame Properties box 236
Table Frame tool 235–236
Table of Contents style properties 140
Table of Contents, in ArcView 46
  WYSIWYG 49
tables, used in maps 235
tabloid size 223, 303
tangents 171
temperature, color ramp for 61
Template feature (Layout) 223
Template Manager 244–245
text 121–166
  changing case 152
  converted to graphic object 151
  copying formatted text 162
  correcting 152
  creating for map legend 137
  explanatory, in map composition 219
  exporting editable 161
  importing 244
  labeling process in ArcView 135–165
  labeling systems 128
  positioning guidelines 130–135, 149
  removing underscores 152
  uses of 127–130
Text and Label defaults (Graphics) 154
Text and Label Defaults dialog box 143
text color 100
text files, importing with Mapper extension 271
Text Label preferences 156
Text Labels properties box 149
Text Properties dialog box 144, 152, 244
Text tool 141–146
  drop-down palette 142, 144
  icon 142
  using 142
Text tool (Layout) 243
Text tool (Mapper extension) 271
texture variables, defined 24
thematic maps 5
  choropleth 50
  dot maps 51
  drafts 40
  isarithmic 51
  types of 50
theme attribute tables
  fields used for labeling 148–153
  linked to text 152
theme layering
  adjusting 37
  for contrast 28

  for hierarchical organization 29, 34
  guidelines for 46
Theme Properties dialog box 138
themes
  changing projections in 192
  labeling 151
  precautions for copying 47
  renaming for clarity 49, 137–138
  repeating 47
  saving active themes 192
  using multiple copies of 248, 249
  See also theme layering
TIF files 237, 287, 314
  saved from PDF files 344
titles, map 7, 33
  adding 243
  changing 161
  default 161
  in map composition 216
TOC Style dialog box 140, 141
topographic maps 51, 282
transmissive media 12
  color values on 41
  legibility in 42
transp.avp 95
Transverse Mercator projection 180, 196, 200
Trimble Pathfinder 211
Truetype fonts 125
two-/three-color ramps 60, 61
two-color color ramps 60
two-color monochromatic color ramps 61
type
  character spread guidelines 135
  color in 129
  curved or angled 133, 146
  defined 123
  orientation guidelines 134
  overlapping 129
  setting in ArcView 135–166
  size guidelines 155
  size of 130
  styles for features 129
  types of 124
  See also type size
type size
  in annotation themes 146
  selecting 143
typefaces
  common examples 123
  defined 123
typestyles 123
  for margin information 220

in annotation themes 146
in custom legends 264
in TOC Style dialog box 140
selecting 143
typography 121–166
terminology 122–124

U

U.S. Geological Survey maps 293
U.S. Interstate label 153
U.S. Interstate label icon 142
U.S. Route label 153
U.S. Route label icon 142
uneven distribution of data 77
Ungroup (Graphics), using 145
unique value legend type 54, 65–68
Unit drop-down menu 223
units (distance), setting 184, 228
units (map), setting 184, 228
Universal Transverse Mercator (UTM) 177, 194, 197
illustrated example 183
limitations of 182
UNIX, transferring graphics with 336
Unofficial AI and AV Symbol page 111
unprojected data 186
illustrated 187
usgs.avp 95
UTM zones 197
UTM. *See* Universal Transverse Mercator

V

Value column, Legend Editor 59
value orientation, defined 25
value variables, illustrated 23
Values Field, Legend Editor 65
values, color
contrasting 41
defined 101
values, grayscale
contrasting 41
vector data 18
compared with raster data 20
vector files
CGM 336
EPS 334
supported for export 333
vector themes, saving 344
Versatec Electrostatic 312
Vertex Edit tool (View Window) 145

Vertical Near-Side Perspective projection 200
Vertical Spacing (Text Properties) 144
vfe.avx 267
view document
advanced techniques 248–261
using 34
View Frame Extent Nudger 275, 267
See also Nudger extension
View Frame Extent Nudger extension 267
View Frame Properties dialog box 227
View Frame tool 226–229
View Properties
changing projections in 191
custom projections 198–200
Custom radio button 195
projection feature 191–200
View Properties dialog box 192
visual contrast 27, 113
experimenting with 36
visual hierarchy 110
visual variables
defined 22
illustrated 23
VRF files 312

W

water buffers 250–253
water.avp 95
weather.avp 95
web, maps for 345–350
web browsers, viewing maps from 339
web sites, for metadata 15
WGS84. *See* World Geodetic System of 1984
Windows Bitmap 333
Windows Metafile 333
WMF files 237, 333
working directory, setting 234
World Geodetic System of 1984 (WGS84) 176, 206
world-from-space projection 193, 195

X

x, y values 20, 177
XBM files 237
Xtools extension 248, 250

Z

z values, types of 20
zmapper.avx 267

Also Available from OnWord Press

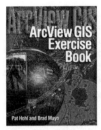

ArcView GIS Exercise Book, Second Edition

Pat Hohl and Brad Mayo

Written to Version 3.x, this book includes exercises on manipulation of views, themes, tables, charts, symbology, layouts and hot links, and real world applications such as generating summary demographic reports and charts for market areas, environmental risk analysis, tracking real estate listings, and customization for task automation.

Order number 1-56690-124-3

480 pages, 7" x 9" softcover

ArcView GIS/Avenue Developer's Guide, Third Edition

Amir Razavi

This books continues to offer readers one of the most complete introductions to Avenue, the programming language of ArcView GIS. By working through the book, intermediate and advanced ArcView GIS users will learn to customize the ArcView GIS interface; create, edit, and test scripts; produce hardcopy maps; and integrate ArcView GIS with other applications.

Order number 1-56690-167-7

432 pages, 7" x 9" softcover

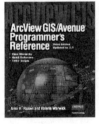

ArcView GIS/Avenue Programmer's Reference, Third Edition

Amir Razavi and Valerie Warwick

This all-new edition of the popular *ArcView GIS/ Avenue Programmer's Reference* has been fully updated based on ArcView GIS 3.1. Included is information on more than 200 Avenue classes, plus 101 ready-to-use Avenue scripts—all organized for optimum accessibility. The class hierarchy reference provides a summary of classes, class requests, instance requests, and enumerations. The Avenue scripts enable readers to accomplish a variety of common customization tasks, including manipulation of views, tables, FThemes, IThemes, VTabs, and FTabs; script management; graphical user interface management; and project production documentation.

Order number 1-56690-170-7

544 pages, 7 3/8" x 9 1/8"

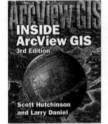

INSIDE ArcView GIS, 3rdEdition

Scott Hutchinson and Larry Daniel

Written for the professional seeking quick proficiency with ArcView, this new edition provides tips on making the transition from earlier versions to the current version, 3.2, and includes an overview of new extensions. The book also presents the software's principal functionality through the development of an application from start to finish, along with several exercises. A companion CD-ROM includes files necessary to follow along with the exercises.

Order number 1-56690-169-3

512 pp., 7-3/8 x 9-1/8" softcover

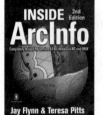

INSIDE ArcInfo, 2E

Jay Flynn and Teresa Pitts

Updated to ArcInfo v. 8 covering the new drag-and-drop interface modules ArcCatalog, ArcMap, and ArcToolbox with a PowerStart exercise to get you making maps right away. Also covers the traditional modules Arc, ArcEdit, and ArcPlot, including tips and exercises. Includes companion CD-ROM.

Order number 1-56690-194-4

492 pages, 7-3/8 x 9-1/4" softcover

ArcView GIS Avenue Scripts: The Disk, Third Edition

Valerie Warwick

All of the scripts from the *ArcView GIS/Avenue Programmer's Reference, Third edition*, with installation notes, ready-to-use on disk. Written to Release 3.1.

Order number 1-56690-171-5

3.5" disk

GIS Data Conversion: Strategies, Techniques and Management

Pat Hohl, Ed.

An in depth orientation to issues involved in GIS data conversion projects, ranging from understanding and locating data, through selecting conversion and input methods, documenting processes, and safeguarding data quality.

Order Number: 1-56690-175-8

432 pages, 7" x 9" softcover

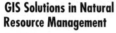

GIS Solutions in Natural Resource Management

Stan Morain

This book outlines the diverse uses of GIS in natural resource management and explores how various data sets are applied to specific areas of study. Case studies illustrate how social and life scientists combine efforts to solve social and political challenges, such as protecting endangered species, preventing famine, managing water and land use, transporting toxic materials, and even locating scenic trails.

Order number 1-56690-146-4

400 pages, 7" x 9" softcover

The GIS Book, 4th Edition

George B. Korte, P.E.

Proven through three highly praised editions, this completely revised and greatly expanded resource is for anyone who needs to understand what a geographic information system is, how it applies to their profession, and what it can do. New and updated topics include trends toward CAD/GIS convergence, the growing field of systems developers, and the latest changes in the GIS landscape.

Order number 1-56690-127-8

440 pages, 7" x 9" softcover

Raster Imagery in Geographic Information Systems

Stan Morain and
Shirley López Baros, editors

This book describes raster data structures and applications. It is a practical guide to how raster imagery is collected, processed, incorporated, and analyzed in vector GIS applications, and includes over 50 case studies using raster imagery in diverse activities.

Order number 1-56690-097-2

560 pages, 7" x 9" softcover

Exploring Spatial Analysis

Yue-Hong Chou

Written for geographic information systems (GIS) professionals and students, this book provides an introduction to spatial analysis concepts and applications. It includes numerous examples, exercises, and illustrations.

Order number 1-56690-119-7

496 pages, 7" x 9" softcover

Processing Digital Images in GIS: A Tutorial Featuring ArcView and ARC/INFO

David L. Verbyla
and Kang-tsung (Karl) Chang

This book is a tutorial on becoming proficient with the use of image data in projects using geographical information systems (GIS). The book's practical, hands-on approach facilitates rapid learning of how to process remotely sensed images, digital orthophotos, digital elevation models, and scanned maps, and how to integrate them with points, lines, and polygon themes. Includes companion CD-ROM.

Order number 1-56690-135-9

312 pages, 7" x 9" softcover

Focus on GIS Component Software: Featuring ESRI's MapObjects

Robert Hartman

This book explains what GIS component technology means for managers and developers. The first half is oriented toward decision makers and technical managers. The second half is oriented toward programmers, illustrated through hands-on tutorials using Visual Basic and ESRI's MapObject product. Includes companion CD-ROM.

Order number 1-56690-136-7

368 pages, 7" x 9" softcover

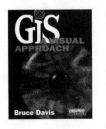

GIS: A Visual Approach

Bruce Davis

This is a comprehensive introduction to the application of GIS concepts. The book's unique layout provides clear, highly intuitive graphics and corresponding concept descriptions on the same or facing pages. It is an ideal general introduction to GIS concepts.

Order number 1-56690-098-0

400 pages, 7" x 9"

GIS: A Visual Approach Graphic Files

This set of 12 disks includes 137 graphic files in Adobe Acrobat, plus the Acrobat Reader. Corresponding with chapters in *GIS: A Visual Approach*, nearly 90% of the book's images are included. Available in Windows or Mac platforms. Ideal for instructors and organizations with a large GIS user base.

Order number 1-56690-120-0

Set of disks

OnWord Press
Thomson Learning™